LEARNING AI FOR BEGINNERS

EASY BITE-SIZED STEPS TO MASTER GENERATIVE AI, MACHINE LEARNING, AND CHATGPT TO ADVANCE YOUR CAREER AND ENHANCE DECISION-MAKING, WITHOUT CODING OR COMPLEXITY

ELI CODERIGHT

CONTENTS

Introduction 5

1. DEMYSTIFYING AI AND ITS EVERYDAY IMPACT 7
   1.1 What Is AI? Breaking Down the Basics 7
   1.2 AI in Your Pocket: Everyday Uses of AI Technologies 11
   1.3 From Sci-Fi to Reality: Real-World AI Applications 14

2. GETTING STARTED WITH USER-FRIENDLY AI TOOLS 19
   2.1 Choosing User-Friendly AI Platforms 19
   2.2 Setting Up Your AI Workspace: Tools and Technologies 23
   2.3 Navigating AI Interfaces: A Beginner's Guide 27

3. UNDERSTANDING DATASETS, YOUR AI GOLDMINE 32
   3.1 Selecting the Right Dataset for Your Project 32
   3.2 Gathering Data: Where to Find Datasets 37
   3.3 Securing and Preparing Your Data for AI 41

4. DATA EXPLORATION AND PREPROCESSING 47
   4.1 Exploring Your Data: Visualization Techniques 47
   4.2 Cleaning and Preprocessing Data for Analysis 52

5. BUILDING YOUR FIRST MODEL 57
   5.1 Choosing the Right AI Model for Beginners 57
   5.2 Setting Up Your First AI Model: A Step-By-Step Guide 62
   5.3 Training Basics: How to Train Your First Model 67

6. EVALUATING AND IMPROVING MODELS 72
   6.1 Measuring Success: Evaluating Model Performance 72
   6.2 Iterative Learning: How to Improve Your Model 77
   6.3 Addressing Overfitting and Underfitting in Models 81

7. MAKING PREDICTIONS AND APPLICATIONS 89
   7.1 From Data to Decision: Making Predictions With AI 89
   7.2 Applying AI Models in Everyday Scenarios 94
   7.3 Creative Applications: Using AI in Novel Ways 99

8. SHARING AND COLLABORATING ON AI PROJECTS 104
8.1 Saving and Exporting Your AI Models 104
8.2 Sharing Your AI Work With the Community 109
8.3 Collaborating on AI Projects: Teamwork in Action 113

9. ETHICAL AI AND RESPONSIBLE USE 119
9.1 Understanding AI Ethics: What You Need to Know 119
9.2 Navigating Privacy and Bias in AI 124
9.3 Implementing Responsible AI Practices 128

10. EXPLORING ADVANCED AI CONCEPTS 134
10.1 Diving Into Neural Networks and Deep Learning 134
10.2 Natural Language Processing: AI and Human Language 139
10.3 The Future of AI: Trends and Innovations 143

11. ENHANCING SKILLS WITH INTERACTIVE EXERCISES 149
11.1 Hands-on Practice: Building a Simple AI Model 149
11.2 Self-Assessment: Testing Your AI Knowledge 151
11.3 Project-Based Learning: Create Your AI Application 153

12. REVISITING AI'S IMPACT IN SEVERAL INDUSTRIES IN MORE DEPTH 156
12.1 AI in Healthcare: Transforming Patient Care 156
12.2 AI in Finance: Revolutionizing Banking and Investments 158
12.3 AI in Entertainment: Changing the Face of Media 160
12.4 The Road Ahead: Embracing Lifelong AI Learning 162

Conclusion 167
Glossary 169
References 183

INTRODUCTION

Have you ever wondered how your favorite music-streaming app knows the right song to play next? Or how your email inbox magically filters out spam and organizes your messages? The answer lies in the fascinating world of artificial intelligence (AI).

AI is no longer a future sci-fi concept. It's now part of our daily lives, making tasks more straightforward, efficient, and enjoyable. From virtual assistants like Siri and Alexa to self-driving cars and personalized shopping recommendations, AI is quietly revolutionizing how we live and work.

But here's the thing: You don't need to be a coding whiz or have a Ph.D. in computer science to understand and even master AI. That's where this book comes in. *Learning AI for Beginners* is your friendly, no-nonsense guide to demystifying AI and putting it to work for you, no matter your background or technical expertise.

I know how intimidating AI can seem at first glance. The jargon, the complex algorithms, and the endless lines of code—it's enough to make anyone's head spin. But trust me—once you break it down into bite-sized pieces and focus on practical applications, AI becomes not only accessible but also incredibly empowering.

Throughout this book, we'll do walk-throughs to help you build fundamental, practical AI skills. We'll start by choosing the right platform for your needs and then dive into selecting and preprocessing datasets. From there, we'll explore different AI models, train them, and evaluate their performance. By the end, you'll be able to make predictions, iterate and improve your models, and even save and share them with others.

But this isn't just a dry, theoretical textbook. We'll focus on real-world examples and hands-on exercises that reinforce your learning and show you how to apply AI in your everyday life and career. Want to automate repetitive tasks at work? Create a chatbot for your website? Predict trends in your industry? AI can help you with all of that and more.

As a bonus, each chapter includes self-assessment quizzes and practical exercises to help cement your understanding and build confidence in your new skills. Plus, there is a handy glossary at the end of the book to help you quickly look up any unfamiliar terms.

However, the most exciting aspect of AI is its potential to transform your career. As more and more industries embrace the technology, the demand for AI-savvy professionals is skyrocketing. Whether you're looking to boost your résumé, transition into a new field, or stay ahead of the curve, learning about AI can open up a world of opportunities.

So, why did I write this book?

Simple: I'm passionate about helping people like you—beginners who might feel daunted by the complexity of AI—not only understand this transformative technology but also feel confident using it to enhance your life and career. I firmly believe that AI shouldn't be reserved for the tech elite; it's a tool everyone should have access to and feel motivated to use.

So, are you ready to demystify AI and unlock its potential?

Let's dive in together. With *Learning AI for Beginners* as your guide, you'll soon be navigating the exciting world of AI with ease and confidence, ready to take your skills to the next level, and seize the opportunities that await. Trust me—by the end of this book, you'll be amazed at what you can achieve with AI—no previous coding skills required.

1

DEMYSTIFYING AI AND ITS EVERYDAY IMPACT

Do you remember the last time you used your smartphone to find the quickest route to your favorite café? Or perhaps when you asked your virtual assistant to set a reminder for a meeting? AI makes these small, everyday conveniences possible.

This chapter discusses breaking down the barriers that make AI seem daunting. We'll explore what AI is and why it matters. By understanding the basics, you can start to see AI not as a complex enigma but as a fascinating and accessible tool.

1.1 WHAT IS AI? BREAKING DOWN THE BASICS

As I've mentioned, you don't need to be a programmer or a mathematician to grasp AI's potential. Think of it as a system **mimicking human thought processes,** helping us solve problems more efficiently. It's all around us, from the algorithms that personalize your news feed to the data-driven models that predict weather patterns.

In other words, AI is creating systems that replicate human intelligence.

This means designing machines that can reason, learn, and make decisions like humans. It might sound like something out of a science fiction novel, but AI is rooted in practical application. The concept has been around for centuries, with early inklings found in the myths of intelligent automata from ancient Greece and mechanical men from ancient China.

Fast-forward to the 20th century, and you'll encounter pivotal moments like the **1956 Dartmouth Conference**, where the term "artificial intelligence" was first coined. Since then, AI has evolved from theoretical musings to tangible innovations that impact every corner of our lives.

One of the most common misconceptions is confusing AI with **machine learning**.

While they are related, they aren't the same. Think of AI as the broader umbrella under which machine learning falls. AI encompasses various technologies that mimic human capabilities, such as reasoning and language processing. On the other hand, machine learning is a specific approach within AI that involves teaching machines to learn from data without being explicitly programmed. This distinction is vital because while AI can function with **structured**, **semistructured**, and **unstructured data**, machine learning primarily deals with structured and semistructured data. We'll get into that in more depth later.

AI's Core Components

The magic of AI lies in its core components: **algorithms**, **data**, and **models**.

- **Algorithms** act as rules or instructions that guide a machine in solving problems or making decisions. They're the brains behind the operations, analyzing data to identify patterns and make predictions.
- **Data** is the lifeblood of AI, feeding these algorithms with the information they need to learn and evolve. Without rich, accurate data, AI systems would be like a painter without a canvas.
- **Models** are the structures built from algorithms and data, forming the blueprint that allows AI to function effectively. A well-constructed AI model can sift through vast amounts of information, extracting valuable insights and making informed predictions.

Many people find AI intimidating because of its perceived complexity and the jargon often associated with it. Terms like "neural networks" and "deep learning" can sound daunting but are actually relatively straightforward. By demystifying these terms, you can see AI as a practical tool rather than a convoluted mystery:

- **Neural networks** are based upon how the human brain's network of neurons operates and are used in AI to recognize patterns and make decisions.
- **Deep learning** is a subset of machine learning involving multiple layers of neural networks, allowing for more complex data processing and pattern recognition.

AI's potential is immense and far-reaching. It's not just about futuristic robots or advanced technologies; it's a transformative force that can enhance everyday experiences. From virtual assistants that help you manage your schedule to recommendation engines that suggest music you'll love, AI is making life more convenient and personalized. It's leveraging technology to improve efficiency and create new opportunities, whether at home, at work, or in the broader world.

Multiple-Choice Questions

1. What is the primary goal of AI?

   a. To replicate human intelligence in machines
   b. To replace humans in all tasks
   c. To create autonomous robots
   d. To process unstructured data exclusively

Answer: a. To replicate human intelligence in machines

2. Which conference is credited with coining the term "artificial intelligence"?

   a. The Turing Conference
   b. The Dartmouth Conference
   c. The AI Summit
   d. The Neural Network Symposium

Answer: b. The Dartmouth Conference

3. How do AI and machine learning differ?

   a. AI is a part of machine learning.
   b. There is no difference; they are the same.
   c. AI only deals with humanlike tasks.
   d. Machine learning is a type of AI that learns from data.

Answer: d. Machine learning is a type of AI that learns from data.

4. What are the essential components of an AI system?

   a. Hardware, software, and networking
   b. Algorithms, data, and models
   c. Programming, output, and input
   d. Users, interfaces, and applications

Answer: b. Algorithms, data, and models

5. What role do algorithms play in AI?

   a. They store large datasets.
   b. They are scripts that execute tasks manually.
   c. They act as the rules or instructions for problem-solving.
   d. They serve as user interfaces for programs.

Answer: c. They act as the rules or instructions for problem-solving.

6. What is meant by "deep learning" in the context of AI?

   a. Learning from simple datasets
   b. Involvement of a single layer of processing
   c. Use of multiple layers of neural networks for complex data processing
   d. Learning that is supervised by humans

Answer: c. Use of multiple layers of neural networks for complex data processing

7. Which of the following is *not* a benefit of AI mentioned in the content?

a. Predicting the future with 100% accuracy
b. Making life more convenient
c. Enhancing everyday experiences
d. Creating new opportunities

Answer: a. Predicting the future with 100% accuracy

1.2 AI IN YOUR POCKET: EVERYDAY USES OF AI TECHNOLOGIES

Imagine waking up to the gentle nudge of your smartphone's virtual assistant. It's already set the perfect temperature in your room and adjusted the lighting to mimic a sunrise, easing you into the day.

This isn't a scene from a futuristic movie; it's the reality of AI working quietly behind the scenes in your everyday life.

Consider your smartphone assistant—maybe Siri or Google Assistant—which is much more than just a voice that sets reminders. These assistants are powered by AI to understand your speech, manage your schedules, and even answer your questions. They learn from your interactions, becoming more personalized over time. It's akin to having a digital concierge who anticipates your needs before you even voice them.

Navigation apps are also a prime example of AI's silent yet tremendous impact. When you use Google Maps or Waze to find the best route to work, AI algorithms process real-time traffic data, predict congestion, and suggest alternatives. It's like having a personal navigator who knows every shortcut and can foresee traffic jams before they happen. These apps save time and reduce stress, demonstrating AI's practical utility in everyday life.

AI doesn't stop at convenience; it also enhances productivity. Consider how AI helps with calendar management. Intelligent scheduling tools can sift through your emails, suggest meeting times, and even book conference rooms, streamlining your day-to-day tasks.

Smart home devices like **Nest thermostats** or **Philips Hue lighting systems** learn your preferences and adjust settings to create the perfect environment without you lifting a finger. These technologies are not just automating tasks but also are creating a seamless living experience.

AI and Communication

AI's role in communication is expanding rapidly. Language-translation apps, for example, break down language barriers by providing real-time translations, giving you the ability to talk effortlessly to people from different parts of the world.

Customer service is being revolutionized by AI-powered chatbots. They can provide instant responses to queries, no matter the time of day. They handle routine inquiries, freeing people to take care of other issues, thus improving the overall efficiency of service delivery.

In the world of entertainment, AI is a game-changer. Streaming services like **Netflix** or **Spotify** use AI to analyze your viewing and listening habits, recommending content customized to your tastes. It's as if they know you better than you know yourself, serving up the next binge-worthy series or album just when you need it.

In the gaming industry, AI creates dynamic environments and adaptive characters that respond to your actions, leading to immersive experiences that keep you engaged.

AI's integration into everyday life is proof of its versatility and potential. From managing your home to transforming your entertainment, AI technologies continue to evolve, making life simpler, more connected, and enjoyable.

Multiple-Choice Questions

1. What is one example of an AI technology that assists you in your daily routine?

   a. A traditional alarm clock
   b. A smartphone's virtual assistant (like Siri or Google Assistant)
   c. A printed calendar
   d. A paper map

Answer: b. A smartphone's virtual assistant (like Siri or Google Assistant)

2. How do navigation apps utilize AI?

a. By providing voice directions without any updates
b. By displaying maps from the 1990s
c. By offering fixed routes to a destination
d. By processing real-time traffic data and suggesting alternative routes

Answer: d. By processing real-time traffic data and suggesting alternative routes

3. What function do intelligent scheduling tools serve?

a. They remind you of your past meetings.
b. They filter spam emails.
c. They help manage your calendar by suggesting meeting times and booking rooms.
d. They organize your personal files.

Answer: c. They help manage your calendar by suggesting meeting times and booking rooms.

4. How do AI-powered chatbots improve customer service?

a. By providing personalized human interaction
b. By offering instant responses to routine inquiries
c. By handling all types of complex issues
d. By eliminating the need for customer service altogether

Answer: b. By offering instant responses to routine inquiries

5. In what way do streaming services like Netflix and Spotify use AI?

a. To analyze your habits and recommend personalized content
b. To load content faster
c. To increase subscription prices based on demand
d. To provide live streaming of events

Answer: a. To analyze your habits and recommend personalized content

6. What is a significant benefit of AI in smart home devices?

a. They require constant manual adjustments.
b. They help create a seamless living experience by learning user preferences.
c. They operate solely on a schedule set by the user.
d. They can only be controlled on smartphones.

Answer: b. They help create a seamless living experience by learning user preferences.

7. Why do some people view AI technologies as impactful in their daily lives?

a. Because they always require manual input
b. Because they increase the complexity of tasks
c. Because they enhance convenience and productivity without users noticing
d. Because they only serve one specific purpose

Answer: c. Because they enhance convenience and productivity without users noticing

1.3 FROM SCI-FI TO REALITY: REAL-WORLD AI APPLICATIONS

Imagine walking into a doctor's office and receiving a diagnosis that is not based just on your symptoms but on a vast array of data that includes your genetic background, lifestyle, and even the air quality of your neighborhood.

This is not futuristic speculation; it's the reality of AI in healthcare. AI-driven diagnostic tools transform medical practices by providing more accurate and timely diagnoses. They analyze complex datasets, uncovering patterns that can easily be missed by human eyes. Individual data can be used to personalize medicine and customizes treatment plans, leading to more effective care with fewer side effects.

AI's role in healthcare extends further, improving administrative processes, streamlining patient management, and even assisting in surgical procedures with precision and accuracy.

Agriculture and Finance

In agriculture, AI is enhancing how we grow and harvest food. Precision farming is becoming crucial with global challenges like climate change and population growth. AI technologies help farmers monitor crop health, predict yields, and optimize resource use. AI analyzes field conditions using sensors and drones, providing actionable insights into soil health and pest infestations, which allows for precise interventions, reducing waste and increasing productivity. As a result, farmers are able to make decisions that boost efficiency and sustainability, ensuring that the demands of a growing global population can be met.

The financial sector is also experiencing a revolution brought on by AI. Fraud-detection systems that learn from vast amounts of transaction data can now identify suspicious activities in real time, safeguarding your finances against cyber threats.

Automated financial advising, or robo-advisors, offers personalized investment strategies based on your personal financial goals and risk tolerance. These intelligent systems make financial planning more accessible and less expensive, opening doors for individuals who might not have considered investing otherwise. AI is reshaping the economic landscape, making it more secure, efficient, and inclusive.

Thanks to AI technologies, transportation is undergoing a significant transformation. On roads all over the world, Self-driving cars, are now being tested. These vehicles use AI to navigate traffic, recognize obstacles, and make split-second decisions to reduce accidents and improve traffic flow.

Traffic-management systems utilize AI to monitor and manage congestion, optimize routes, and reduce travel times. This smart transportation approach enhances safety and reduces emissions and fuel consumption which contributes to environmental sustainability.

Industry and Production

In the fields of industry and production, AI is dramatically improving operational efficiencies. **Predictive maintenance** is one such application, where AI systems can identify possible equipment failures before they occur, minimizing downtime and reducing costs and allowing for timely maintenance.

Quality control is another area where AI shines, using **machine vision** to detect product defects, ensuring that only the best reach consumers. This level of precision and reliability drives innovation and competitiveness in manufacturing, setting new standards for quality and efficiency.

In addition, AI is making strides in areas like **energy management**, where smart grids and predictive analytics are helping optimize resource use and support sustainable practices.

As we look to the future, AI continues to evolve, opening new possibilities and applications. One exciting trend is the development of AI systems that can explain their decision-making processes, known as **explainable AI**. This capability promises to build trust and understanding between humans and machines, enabling broader adoption of AI technologies. The potential of AI is vast, and its applications are limited only by our imagination.

Multiple-Choice Questions

1. How is AI transforming healthcare?

   a. By eliminating the need for doctors
   b. By providing accurate diagnoses using a vast array of data
   c. By focusing solely on administrative tasks
   d. By replacing medical records with paper forms

Answer: b. By providing accurate diagnoses using a vast array of data

2. What is "precision farming" in the context of AI?

   a. Farming that relies heavily on traditional methods
   b. A method to grow crops without data analysis
   c. Using AI to monitor crop health and optimize resource use

d. Farming that disregards environmental considerations

Answer: c. Using AI to monitor crop health and optimize resource use

3. How does AI aid in the financial sector?

a. By making physical bank tellers obsolete
b. Through implementing lengthy paperwork for financial planning
c. By enhancing fraud detection and providing automated financial advice
d. By discouraging investments in stocks

Answer: c. By enhancing fraud detection and providing automated financial advice

4. What does the term "robo-advisors" refer to in finance?

a. Robots that physically manage money
b. AI systems that offer personalized investment strategies
c. Human financial advisors who use AI tools
d. Automated teller machines

Answer: b. AI systems that offer personalized investment strategies

5.. What significant advancements are being made in transportation due to AI?

a. Development of self-driving cars and improved traffic-management systems
b. Increased reliance on paper maps
c. Introduction of horse-drawn carriages on highways
d. Banning all forms of automation in vehicles

Answer: a. Development of self-driving cars and improved traffic management systems

6. In manufacturing, how is AI enhancing operational efficiency?

a. By increasing the number of manual inspections required on the production line
b. By reducing the use of technology in quality control
c. By completely automating factory work without human oversight
d. Through predictive maintenance that anticipates equipment failures

Answer: d. Through predictive maintenance that anticipates equipment failures

7. What is explained by the term "explainable AI"?

a. AI that can explain its decision-making processes to users
b. AI systems that operate without any human input
c. AI that functions only in a research setting
d. AI systems that are difficult to understand

Answer: a. AI that can explain its decision-making processes to users

8. What potential future applications of AI are mentioned in the content?

a. The introduction of more manual labor in industries
b. Developments in energy management through smart grids
c. Reducing the role of technology in daily tasks
d. A regression to preindustrial methods of production

Answer: b. Developments in energy management through smart grids

Now that we've demystified AI, in the next chapter, we'll discuss how you can get started with AI tools, right from the very basics.

2

GETTING STARTED WITH USER-FRIENDLY AI TOOLS

Looking at the world of AI tools can feel like you're standing at the entrance of a vast library filled with endless books, each holding the potential to unlock new worlds. The options might seem overwhelming initially, but you'll soon find the perfect platform that suits your needs.

We're about to explore some of the most user-friendly AI platforms designed for beginners. These platforms make it easy to learn AI without getting overloaded by technical complexities, ensuring a smooth and enjoyable learning experience.

2.1 CHOOSING USER-FRIENDLY AI PLATFORMS

Choosing the right AI platform is like picking the perfect toolkit for a new hobby. You want tools that are not only effective but also accessible and easy to use.

- One popular choice is **Google Colab**. It's a cloud-based environment that offers free access to powerful **GPU resources**, making it ideal for running data-intensive tasks. With Google Colab, you don't have to worry about installing heavy software on your computer; everything **runs smoothly from your web browser**. This platform is particularly attractive for beginners who want to experiment with AI models without the hassle of complex setups. Plus, it's part of the Google ecosystem, which means you get seamless integration with other Google services.

- Another beginner-friendly option is **Microsoft Azure ML Studio**. This platform stands out with its intuitive drag-and-drop interface, allowing you to build and test AI models **without writing a single line of code**. It's perfect for visual learners who prefer interacting with software through a **graphical interface** rather than text-based commands.
- **Azure ML Studio** also provides a range of preconfigured modules, making it easy to create machine learning pipelines. This means you can focus more on understanding the concepts rather than getting stuck on technical details.
- **IBM Watson Studio** offers a suite of prebuilt AI tools that cater to quick implementation. It's like having a set of ready-made solutions at your fingertips, which you can customize to fit your specific needs. Watson Studio excels in **facilitating collaboration**, allowing multiple users to work on projects simultaneously. This feature is particularly useful if you're part of a team or learning group. It's a platform that encourages sharing and collective problem-solving, which can significantly enhance your learning experience.

When selecting an AI platform, consider a few key factors:

- **Cost and accessibility** are crucial; some platforms offer free tiers, while others require a subscription. Google Colab, for instance, provides a generous free version, while Azure offers a pay-as-you-go model.
- **Community support** is another crucial aspect. Platforms with active forums and plentiful tutorials can be invaluable when encountering issues or wanting to expand your skills. On a good platform, you'll have a community of fellow learners and experts ready to help at any moment.
- Evaluating **platform features** is also essential. Look for **integration capabilities**, which determine how effectively a platform can work with various data sources and APIs.

- **A user-friendly interface** is also a must. You want a platform that's easy to navigate so you can focus on learning rather than figuring out how to use the tool itself. For example, Azure's drag-and-drop features make it easy to visualize your data flow, while Colab's code cells encourage interactive coding.

To make an informed decision, be sure the platform's capabilities can help you with your **personal project goals**.

If you aim to build a **small-scale model**, a platform like Google Colab might be perfect. However, if you envision scaling your projects in the future, a more robust solution like Azure or IBM Watson might be more suitable. Also, consider **scalability**—choosing a platform that can grow with your skills ensures you won't outgrow its capabilities too quickly.

Multiple-Choice Questions

1. What is the purpose of a user-friendly AI platform for beginners?

   a. To overwhelm users with technical jargon
   b. To focus solely on coding skills
   c. To replace traditional textbooks
   d. To make learning AI accessible and enjoyable without technical complexities

Answer: d. To make learning AI accessible and enjoyable without technical complexities

2. Which AI platform is known for its cloud-based environment offering free access to powerful GPU resources?

   a. Google Colab
   b. IBM Watson Studio
   c. Microsoft Azure ML Studio
   d. Amazon SageMaker

Answer: a. Google Colab

3. What feature distinguishes Microsoft Azure ML Studio from other platforms?

a. It requires extensive coding knowledge.
b. It has a drag-and-drop interface for building AI models.
c. It only supports large-scale enterprise solutions.
d. It offers no prebuilt modules or tools.

Answer: b. It has a drag-and-drop interface for building AI models.

4. How does IBM Watson Studio encourage collaboration among users?

a. By limiting access to one user at a time
b. By providing prebuilt AI tools that cannot be customized
c. By allowing multiple users to work on projects simultaneously
d. By focusing solely on personal projects without team support

Answer: c. By allowing multiple users to work on projects simultaneously

5. When evaluating an AI platform, which factor is *not* mentioned in the text?

a. Cost and accessibility
b. Availability of mobile applications
c. Community support
d. Integration capabilities

Answer: b. Availability of mobile applications

6. Why is community support important when selecting an AI platform?

a. It guarantees you will not encounter any issues.
b. It replaces the need for comprehensive documentation.
c. It provides coding lessons.
d. Active forums and tutorials can help troubleshoot and expand skills.

Answer: d. Active forums and tutorials can help troubleshoot and expand skills.

7. What should you consider regarding scalability when selecting an AI platform?

a. Selecting a platform that can accommodate your growing skills and project needs
b. Choosing a platform that limits your project size
c. Opting for a platform with minimal features for simplicity
d. Prioritizing platforms that only offer basic functions

Answer: a. Selecting a platform that can accommodate your growing skills and project needs

8. When selecting an AI platform, what is a quick decision-making checklist item?

a. The aesthetic design of the user interface
b. Whether the platform is supported by a celebrity
c. Project goals alignment, including potential scalability
d. The size of the platform's advertising budget

Answer: c. Project goals alignment, including potential scalability

2.2 SETTING UP YOUR AI WORKSPACE: TOOLS AND TECHNOLOGIES

Let's say you're setting up a cozy corner in your home for a new hobby. You'd want it to have the right tools, be organized for efficiency, and perhaps even be inspiring. Your AI workspace should feel the same way, welcoming you to explore and create without hurdles.

- Let's start with the backbone of many AI projects: **Python**. This programming language is noted for simplicity and readability. This makes it a great choice for beginners. Python is the Swiss army knife of coding—versatile and powerful, yet easy to pick up. It supports numerous libraries and frameworks specifically designed for AI. **TensorFlow** and **PyTorch**, are excellent examples that can help streamline complex processes.

- To complement Python, **Jupyter Notebooks** acts as your interactive lab. Think of it as a dynamic notebook where you can document your code, run it in real time, and visualize the results in one place. This tool is incredibly user-friendly and encourages experimentation. You can write notes, insert code blocks, and see your outputs immediately, making it easier to understand what your code is doing and why. Jupyter Notebooks is especially handy for data analysis and model training, allowing you to annotate your thought process and share your work with others.

Installing AI Tools

Installing these tools might sound daunting, but it's more straightforward than you'd expect.

Anaconda Distribution simplifies the process by bundling Python with essential data science libraries all in one go, like a starter kit ready to use right out of the box. Once you download and install Anaconda, you can launch Jupyter Notebooks with just a few clicks. This setup saves time and reduces the chances of running into compatibility issues when installing libraries individually.

Another crucial aspect of setting up your AI workspace is creating **virtual environments**. These are isolated spaces within your system where you can install various versions of libraries and dependencies without affecting other projects. Think of it as having a personalized toolkit for each project, ensuring that updates or changes in one environment don't disrupt another.

Setting up a virtual environment is straightforward with Anaconda; you can create one using simple commands in the terminal, allowing you to switch between projects effortlessly.

A **clean workspace** is critical for maintaining productivity. Start by organizing your digital files systematically. **Create folders for datasets, scripts**, and **documentation** using clear and descriptive names. This organization helps you quickly locate files and ensures your workspace remains clutter-free.

On the **hardware** side, ensure your computer meets the minimum requirements for running basic AI applications. While you don't need an ultra-powerful machine, having a decent processor and sufficient RAM will keep your projects running smoothly.

Collaborative Tools

Collaboration is an integral part of the AI process.

- **GitHub** is one tool that allows you to track changes, collaborate on projects, and share your code with others. The platform acts as a version control system, documenting every change made to your code and enabling you to go back to previous versions if needed. GitHub is a great place to learn from others, as many developers open-source their projects for the community to explore and contribute.
- Platforms like **Slack** or **Discord** are invaluable for communication and community support. These channels let you join groups of like-minded individuals, offering spaces where you can ask questions, share insights, and get feedback. Engaging with a community enriches your learning experience, providing diverse perspectives and creating a sense of belonging in the wide world of AI.

Multiple-Choice Questions

1. What is considered the backbone of many AI projects?

a. Java
b. C++
c. Python
d. R

Answer: c. Python

2. What is Jupyter Notebooks primarily used for in AI projects?

a. Compiling code into applications
b. Documenting code, running it in real time, and visualizing results

c. Storing large datasets securely
d. Conducting team meetings

Answer: b. Documenting code, running it in real time, and visualizing results

3. How does Anaconda Distribution simplify the setup process for an AI workspace?

a. It bundles Python with essential data science libraries.
b. It requires separate installations for each library.
c. It eliminates the need for any coding.
d. It automatically updates all libraries without consent.

Answer: a. It bundles Python with essential data science libraries.

4. What is the purpose of creating virtual environments in an AI workspace?

a. To track changes in software versions across projects
b. To develop hardware solutions
c. To create a graphical user interface for coding
d. To isolate projects so that one project's dependencies do not affect another

Answer: d. To isolate projects so that one project's dependencies do not affect another

5. Why is organizing your digital files important in an AI workspace?

a. It makes the computer run faster.
b. It improves the aesthetics of your screen.
c. It allows you to create a backup automatically.
d. It helps you quickly locate files and keeps the workspace clutter-free.

Answer: d. It helps you quickly locate files and keeps the workspace clutter-free.

6. What feature does GitHub provide for collaborating on AI projects?

a. It offers free cloud storage for all files.
b. It acts as a version-control system.
c. It runs AI models automatically.
d. It compiles projects into standalone applications.

Answer: b. It acts as a version-control system.

7. Which platforms are mentioned as valuable for communication and community support?

a. Slack and Discord
b. Reddit and Quora
c. LinkedIn and Facebook
d. Zoom and Skype

Answer: a. Slack and Discord

8. What is a recommended hardware requirement for an effective AI workspace?

a. A gaming console
b. An ultra-powerful machine is a must
c. A decent processor and sufficient RAM
d. A tablet for portability only

Answer: c. A decent processor and sufficient RAM

2.3 NAVIGATING AI INTERFACES: A BEGINNER'S GUIDE

Think about the last time you sat in a new car. You likely faced a dashboard full of indicators and controls, each with its own purpose, and you learned what each one was for. Navigating AI interfaces is quite similar, but once you familiarize yourself with the layout, you'll find it's a powerful tool at your fingertips rather than a daunting console.

Let's start with the **dashboard**, which is the heart of any AI platform. This is where you'll find an overview of your projects, datasets, and tools, all organized to keep everything you need within reach.

Think of it as your command center, where each section serves a specific function, whether it's managing your projects, accessing data, or tweaking settings. The key is to explore each section by clicking through icons, tabs, and dropdowns. This exploration will help you understand what each part of the interface does, making your workflow more efficient as you become more familiar with the environment.

As you dive deeper into these interfaces, you'll encounter interactive elements designed to encourage learning and experimentation.

In platforms like Jupyter Notebooks, you'll work with **code cells**, which are blocks where you can write, edit, and execute code. This setup allows you to view the results of your code in real time, making it an excellent way to learn through doing. You can run a cell to execute its code, modify it if needed, and organize these cells to structure your projects logically.

Visualization tools are another key feature, enabling you to interpret data through graphs and charts. These visual aids are invaluable for understanding trends and patterns within your datasets. By interacting with these elements, you gain the hands-on experience you need to master AI tools.

Troubleshooting and Debugging

Of course, no journey into AI is without its hiccups.

As you work through projects, you might encounter errors or unexpected results. This is where **troubleshooting** becomes essential. Start by familiarizing yourself with **debugging** basics.

Most coding errors are small—perhaps a misplaced comma or a misspelled variable name—but they can halt your progress. Learning to identify these common mistakes quickly will save you time and frustration.

Additionally, many platforms offer FAQs or forums where you can find solutions to frequent queries. These resources are like having a friendly guide ready to help you navigate rough patches.

Remember, encountering errors is a normal part of learning AI, and each one presents an opportunity to enrich your understanding.

Consider incorporating **shortcuts and productivity hacks** into your routine to boost your productivity and streamline your workflow. Small changes can make a significant difference. For example, familiarize yourself with **keyboard shortcuts** that allow you to quickly work through your code, run cells, or access tools. These shortcuts reduce the time spent on repetitive tasks. This allows you to focus more on the creative aspects of your projects.

Additionally, many platforms offer **customization options**, letting you create the interface to suit your preferences. Whether changing the theme for better visibility or rearranging panels to match your workflow, these tweaks can enhance your overall experience. Embracing these productivity hacks will improve your efficiency and make your interactions with AI platforms more enjoyable and intuitive.

Multiple-Choice Questions

1. What analogy is used to describe the experience of navigating an AI interface?

a. Using a map in an unfamiliar city
b. Reading a book in a foreign language
c. Stepping into a new car for the first time
d. Climbing a mountain

Answer: c. Stepping into a new car for the first time

2. What is the purpose of the dashboard in an AI platform?

a. To run simulations and experiments
b. To provide an overview of projects, datasets, and tools
c. To create marketing materials
d. To track internet usage

Answer: b. To provide an overview of projects, datasets, and tools

3. In Jupyter Notebooks, what are code cells used for?

a. To write, edit, and execute code
b. To store only textual information
c. To display documentation for projects
d. To compile projects into applications

Answer: a. To write, edit, and execute code

4. How do visualization tools aid in working with AI data?

a. They replace the need for code.
b. They are only useful for marketing purposes.
c. They limit the types of data that can be analyzed.
d. They provide a way to interpret data through graphs and charts.

Answer: d. They provide a way to interpret data through graphs and charts.

5. What is a common cause of coding errors mentioned in the text?

a. Complex algorithms
b. Hardware malfunctions
c. Small mistakes, like misplaced commas or misspelled variable names
d. External software conflicts

Answer: c. Small mistakes, like misplaced commas or misspelled variable names

6. What should a beginner do when encountering errors in their code?

a. Ignore the errors.
b. Stop using the platform altogether.
c. Ask a friend to write the code for them.
d. Familiarize themselves with debugging basics and available resources.

Answer: d. Familiarize themselves with debugging basics and available resources.

7. How can shortcuts improve productivity when using AI platforms?

a. They complicate the coding process.
b. They allow for quicker navigation and task completion.
c. They require extensive training to master.
d. They eliminate the need for any coding knowledge.

Answer: b. They allow for quicker navigation and task completion.

8. What customization options might AI platforms offer to enhance user experience?

a. Rearranging panels and changing the theme for better visibility
b. Changing the hardware used for coding
c. Disabling all interactive features
d. Preventing users from modifying the interface

Answer: a. Rearranging panels and changing the theme for better visibility

You now know the basics of AI and what you can achieve with it. Up next, we tackle datasets and how you can use them to make your projects flourish.

3

UNDERSTANDING DATASETS, YOUR AI GOLDMINE

Have you ever lined up all the ingredients to cook a new recipe, ready to make a delicious meal, only to discover that you accidentally swapped sugar and salt? Or grabbed spoiled milk? The outcome would be far from appetizing!

The success of your AI model largely hinges on the quality and relevance of the data you feed it. Just as a recipe relies on the right mix of ingredients to create a dish, an AI project requires a well-chosen dataset to produce accurate and meaningful results.

3.1 SELECTING THE RIGHT DATASET FOR YOUR PROJECT

Choosing a relevant dataset is crucial because the effectiveness and accuracy of your AI project will be directly impacted.

Think of your dataset as the foundation of your model; if it's shaky, the entire structure can falter. Knowing your project goals and choosing a dataset that aligns with them ensures that your model receives the right information to learn and make predictions.

For instance, if you're developing a model to predict house prices, using a dataset that includes relevant variables such as location, size, and number of bedrooms is essential. On the other hand, irrelevant data can lead to inaccurate predictions, making your model less effective.

The importance of **choosing the right dataset** cannot be overstated.

Take the example of a retail company aiming to improve its recommendation system. Initially, they used a dataset focused solely on past purchase history, but the recommendations were not as effective as they'd hoped.

By **expanding the dataset** to include browsing history and demographic information, the company significantly enhanced the system's accuracy, leading to better customer satisfaction and increased sales.

Choosing the Right Dataset for You

When selecting a dataset, several criteria can help you determine its suitability:

- **Data quality indicators**, such as accuracy, completeness, and consistency, play a vital role.
- **Accurate data** reflects real-world scenarios, while **completeness** ensures that all necessary information is present.
- **Consistency** means that the data maintains a uniform format and structure, which is crucial for reliable analysis.
- Additionally, the **size and diversity** of your dataset matter. A larger dataset with diverse data points offers more opportunities for your model to learn and adapt, leading to better performance.
- **Ethical considerations** are also crucial when choosing a dataset.
- **Privacy** concerns should not be overlooked, especially if your data contains sensitive information. Handling such data responsibly is essential to protect individuals' rights and comply with legal regulations.
- Moreover, **bias detection** is necessary to ensure that your data doesn't favor one group over another, leading to unfair outcomes. By identifying and mitigating biases, you promote fairness and equity in your AI models, enhancing their reliability and trustworthiness.

To aid in the decision-making process, consider utilizing visual tools like **decision trees**, which can simplify the selection of a suitable dataset. These tools help map out different pathways and outcomes, providing a clear overview of your options.

Additionally, using a checklist of questions can guide you in evaluating datasets.

Questions such as "Does the dataset match my project goals?" and "Is the data accurate and complete?" can help you make informed decisions. By systematically assessing these factors, you can select a dataset that best supports your project's objectives.

Dataset Selection Checklist

Does the dataset align with the project's goals and objectives?
Are there sufficient data points and diversity to ensure model accuracy?
Is the data accurate, complete, and consistent across entries?
Have privacy concerns and biases been addressed and mitigated?

Approaching dataset selection with a structured framework ensures that your chosen data will be a solid foundation for your AI project. You set the stage for a successful and impactful AI model by carefully considering relevance, quality, and ethics.

Multiple-Choice Questions

1. What analogy was used to explain the importance of selecting the right dataset?

a. Cooking a new recipe
b. Choosing the right kitchen utensils
c. Building a house
d. Painting a picture

Answer: a. Cooking a new recipe

2. Why is choosing a relevant dataset crucial for an AI project?

a. It affects the aesthetics of the model.
b. It directly impacts the effectiveness and accuracy of the model.

c. It simplifies the coding process.
d. It increases the cost of the project.

Answer: b. It directly impacts the effectiveness and accuracy of the model.

3. What should be included in a dataset for a model predicting house prices?

a. Only the historical price data
b. Irrelevant variables like the color of the house
c. Relevant variables like location, size, and number of bedrooms
d. Only customer testimonials

Answer: c. Relevant variables like location, size, and number of bedrooms

4. What data quality indicators are highlighted as important for dataset selection?

a. Accuracy, completeness, and consistency
b. Price, popularity, and availability
c. Size, color, and shape
d. Source, age, and style

Answer: a. Accuracy, completeness, and consistency

5. What role does dataset size and diversity play in AI projects?

a. Larger and more diverse datasets provide better learning opportunities.
b. Both are irrelevant to model performance.
c. Smaller datasets are always more effective.
d. Diversity hinders model performance.

Answer: a. Larger and more diverse datasets provide better learning opportunities.

6. What ethical consideration is crucial when selecting a dataset?

a. The reputation of the data provider
b. Cost of data acquisition
c. Privacy concerns and bias detection
d. The design of the data visualization

Answer: c. Privacy concerns and bias detection

7. What tool is suggested to help in the dataset selection process?

a. Spreadsheets
b. Decision trees
c. Flowcharts
d. Graphs

Answer: b. Decision trees

8. Which of the following is *not* a question from the dataset selection checklist?

a. Does the dataset align with the project's goals?
b. Is the data accurate and complete?
c. Have privacy concerns been addressed?
d. Is the data obtained from a popular source?

Answer: d. Is the data obtained from a popular source?

9. What is the main benefit of expanding a dataset beyond initial parameters, as illustrated by the retail company's example?

a. Increased data collection costs
b. Enhanced system accuracy and customer satisfaction
c. Simplified data analysis
d. Easier data entry

Answer: b. Enhanced system accuracy and customer satisfaction

10. How does a structured approach to dataset selection contribute to an AI project?

a. It complicates the decision-making process.
b. It ensures that the data does not align with project goals.
c. It leads to a weak foundation for the AI model.
d. It sets the stage for a successful and effective AI model.

Answer: d. It sets the stage for a successful and effective AI model.

3.2 GATHERING DATA: WHERE TO FIND DATASETS

When you're ready to gather data for your AI project, you'll find that the world is brimming with opportunities to access free datasets. These resources are like treasure troves, offering a wealth of information that can be pivotal for your project:

- **Government databases**, for example, are a fantastic starting point. National statistical offices often publish public sector information, providing datasets ranging from economic indicators to public health statistics. Using these resources, you can access reliable data collected and maintained by official agencies.
- Similarly, **academic sources** are invaluable, with universities frequently releasing research datasets and publications. These are often peer-reviewed, ensuring a level of credibility that's hard to match.
- **Kaggle** stands out as another popular hub for finding datasets. Known for its vibrant community and machine learning competitions, Kaggle provides a platform where you can access diverse datasets and engage with a community of data enthusiasts. Think of it as a social media site for data scientists, where you can find datasets on anything from global demographics to niche market trends. Not only can you download data, but you can also see how others are using it, offering inspiration and guidance for your projects.

Paying for Datasets

While free datasets are abundant, sometimes your project might require proprietary data. Accessing these datasets often involves subscription services or purchasing licenses. Some services offer tiered access, allowing you to pay for the exact data you need without overspending.

Licensing agreements are key here, as they dictate how to use the data and any restrictions involved. It's crucial to read and understand these terms—they can impact the scope and legality of your work.

Developing relationships with data providers can also be beneficial, as they might offer insights or additional resources that enhance your project.

Web Scraping

Another fascinating way to gather data is through **web scraping**.

This technique involves extracting information from websites, like picking the best fruit from a digital orchard. However, it's vital to practice ethical scraping, ensuring compliance with legal and ethical standards.

Websites often have terms of service that detail what is permissible, and respecting these guidelines is crucial.

Beautiful Soup and Selenium are commonly used web scraping tools:

- **Beautiful Soup** is a library in Python that allows you to navigate and search through HTML and XML files.
- **Selenium** is a more robust tool that can automate web browsing and interactions, making it a powerful choice for complex scraping tasks.

As you venture into data collection, expect to face a few challenges.

Incomplete data is a common hurdle, but there are ways to manage it:

- Techniques like **imputation**, which involves filling in missing values, can help maintain the integrity of your dataset. When dealing with **dynamic data** sources that change over time, you'll need strategies to ensure your data remains up-to-date.

- Setting up **automated scripts** that periodically check for updates can be a lifesaver, ensuring your dataset is as fresh as possible.
- Another issue to watch out for is the **volume of data**. Sometimes, you'll have more data than you know what to do with, which can overwhelm your system.

Structuring your data collection to focus on relevant subsets can help you manage this challenge, making your dataset not only manageable but also more targeted for your specific AI model needs.

Multiple-Choice Questions

1. What is the primary advantage of using government databases for data collection?

   a. They provide free access to datasets maintained by official agencies.
   b. They are the only source of high-quality data.
   c. They require multiple subscriptions for access.
   d. They contain outdated information.

Answer: a. They provide free access to datasets maintained by official agencies.

2. Which platform is highlighted as a community hub for accessing datasets and engaging with other data enthusiasts?

   a. GitHub
   b. Google Scholar
   c. Kaggle
   d. Wikipedia

Answer: c. Kaggle

3. What is a key characteristic of datasets released by academic sources?

   a. They often contain irrelevant information.
   b. They are not typically peer-reviewed.
   c. They typically possess a high level of credibility.

d. They require special skills to access.

Answer: c. They typically possess a high level of credibility.

4. When you need proprietary data for your project, what is often required?

a. Accessing government databases
b. Entering community competitions
c. Collaborating with academic institutions
d. Purchasing licenses or subscription services

Answer: d. Purchasing licenses or subscription service

5. What is the purpose of web scraping in data collection?

a. To extract information from websites
b. To automate the analysis of existing datasets
c. To gather data from structured databases only
d. To summarize data from multiple sources

Answer: a. To extract information from websites

6. Which of the following tools is commonly used for web scraping?

a. TensorFlow
b. Beautiful Soup
c. Matplotlib
d. Scikit-learn

Answer: b. Beautiful Soup

7. What is a common challenge faced during data collection?

a. Excessive reliability of sources
b. Incomplete data
c. Finding datasets that are too affordable
d. Surplus of academic publications

Answer: b. Incomplete data

8. What is the technique called that involves filling in missing values in a dataset?

   a. Imputation
   b. Normalization
   c. Compression
   d. Augmentation

Answer: a. Imputation

9. What strategy can be utilized to keep dynamic data sources up-to-date?

   a. Manual updates every month
   b. Deleting outdated data
   c. Setting up automated scripts to check for updates
   d. Ignoring the changes completely

Answer: c. Setting up automated scripts to check for updates

10. What is a good practice for effective data management when dealing with a large volume of data?

   a. Collect more data indiscriminately.
   b. Skip data collection altogether.
   c. Store all data in a single file.
   d. Focus on relevant subsets of data.

Answer: d. Focus on relevant subsets of data.

3.3 SECURING AND PREPARING YOUR DATA FOR AI

Think of your data as a treasure chest filled with valuable insights. Just as you would secure a chest with a lock and key, your data needs protection to ensure its integrity and privacy:

- **Data security** keeps your information safe and helps you maintain trust and credibility in your AI projects.

- **Encryption** is one method to safeguard data both in transit and at rest. This technique involves converting your data into a coded format that can only be accessed by those with the correct decryption key. It's like speaking in a private language that only trusted parties understand.
- **Access control** is another crucial aspect, allowing you to manage who can view or modify your datasets. Setting specific permissions ensures that only authorized people can make changes. This reduces the risk of data breaches or unintended alterations.

Once your data is secure, the next step is to prepare it for AI modeling. Methods like data cleaning, normalization, and standardization are common practices here, ensuring that your data fits within a specific scale:

- Think of **data cleaning** as tidying up your workspace before starting a project. It's an essential process that involves identifying and removing duplicates, which are redundant entries that can skew your results. Imagine sorting through a stack of documents and finding several copies of the same page; you would only keep one to avoid clutter.
- Similarly, **normalization** involves standardizing data formats for consistency, ensuring all entries follow the same structure. This step is akin to uniformly organizing your files, making locating and analyzing information easier.
- **Reducing noise** involves filtering out irrelevant data that might distract from meaningful patterns. By eliminating these unnecessary details, you focus on the core elements that truly matter for your analysis.
- **Data transformation** is reshaping your data to fit the AI model's requirements.
- **Feature scaling**, for example, adjusts the range of your data values to prevent any one feature from overshadowing others. Think of it as leveling the playing field, where each variable contributes equally to the model's outcome.

- **Encoding categorical variables** is another transformation technique where nonnumerical data is converted into a format that AI models can process. **One-hot encoding** is a popular method, turning categories into binary vectors. Picture turning a list of fruit names into a matrix where each row represents a fruit, and each column stands for a specific type, marked with a 1 or 0.

Specialized Tools as Digital Assistants

Consider using specialized tools and libraries designed for data management to streamline your data preparation process. Software like Pandas and NumPy are invaluable for handling large datasets, offering functions that simplify data cleaning and transformation tasks:

- **Pandas** enables you to easily manipulate data structures.
- **NumPy** supports mathematical operations.

These tools act as your digital assistants, taking on complex tasks and freeing you to focus on analysis and insight generation. Exploring these resources can significantly enhance your efficiency and accuracy, ensuring your data is in prime condition for AI modeling.

As you continue to navigate the world of AI, remember that securing and preparing your data is not just a preliminary step but a continuous practice. Each project might require different levels of security and preparation, making it essential to adapt and refine your procedures as needed.

Multiple-Choice Questions

1. Why is data security important in AI projects?

   a. It increases the cost of data collection.
   b. It eliminates the need for data preparation.
   c. It simplifies the modeling process.
   d. It is necessary for maintaining trust and credibility.

Answer: d. It is necessary for maintaining trust and credibility.

2. What method is described as converting data into a coded format?

a. Access control
   a. Data cleaning
   b. Encoding
   c. Encryption

Answer: d. Encryption

3. What does access control allow you to manage regarding your datasets?

a. The storage location of your data
b. The format of your data
c. Who can view or modify the datasets
d. The method of data analysis

Answer: c. Who can view or modify the datasets

4. What is the purpose of data cleaning?

a. To recover lost data
b. To prepare the data for secure storage
c. To identify and remove duplicates and irrelevant entries
d. To increase the volume of data

Answer: c. To identify and remove duplicates and irrelevant entries

5. How is normalization related to data preparation?

a. It involves adding new data points to the dataset.
b. It standardizes data formats for consistency.
c. It filters out noise and irrelevant information.
d. It encrypts the data for security.

Answer: b. It standardizes data formats for consistency.

6. What is feature scaling?

a. The adjustment of the range of data values to ensure equitable contribution from all features
b. A process of reducing the dataset size

c. A method of cleaning textual data entries
d. A way to delete unnecessary data points

Answer: a. The adjustment of the range of data values to ensure equitable contribution from all features

7. Which encoding technique converts nonnumerical data into a format that AI models can process?

a. Data cleaning
b. Normalization
c. One-hot encoding
d. Data transformation

Answer: c. One-hot encoding

8. What software tool is mentioned as helpful for handling large datasets?

a. Pandas
b. Tableau
c. Microsoft Word
d. PowerPoint

Answer: a. Pandas

9. Why is data transformation necessary before applying AI models?

a. To store data in a cloud environment
b. To ensure the data format meets the model's requirements
c. To visualize the data in charts
d. To simplify the overall AI project

Answer: b. To ensure the data format meets the model's requirements

10. What is the nature of securing and preparing data for AI?

a. It is a one-time task that can be ignored in future projects.
b. It is a preliminary step that should be done quickly.

c. It is an ongoing practice that may need adaptation with each project.
d. It is unnecessary if you use proprietary data sources.

Answer: c. It is an ongoing practice that may need adaptation with each project.

As we wrap up this chapter, reflect on your journey through the world of datasets. You've learned how to select the right dataset, where to find them, and how to prepare your data for analysis. These skills are fundamental, forming the backbone of any successful AI project.

Embrace the process, celebrate your progress, and look forward to the exciting developments that lie ahead. As you move into the next chapter, you'll discover how these foundational skills come to life in creating and training AI models, opening new doors to innovation and discovery.

4

DATA EXPLORATION AND PREPROCESSING

Picture yourself as a detective, sifting through a complex web of clues to solve a mystery. In the world of data, visualization is your magnifying glass, helping you uncover hidden patterns and insights that might otherwise remain elusive.

Data visualization is not merely about making charts look pretty; it's an essential tool for making complex datasets more digestible and understandable. Translating raw data into visual formats lets you easily **identify trends, spot anomalies, and make informed decisions**.

This chapter will guide you through the art of data visualization, introducing you to various tools and techniques that can bring your data to life, helping you make sense of the numbers and see the bigger picture.

4.1 EXPLORING YOUR DATA: VISUALIZATION TECHNIQUES

Visualization transforms complex data into a language that's easier to understand. Just as a picture is worth a thousand words, a graph or chart can reveal stories hidden in rows of numbers.

- For instance, a **scatter plot** can illustrate relationships between

variables, such as how sales figures might correlate with marketing spend.
- Similarly, **histograms** can show the distribution of a dataset, highlighting how data points are spread across different ranges.
- **Heat maps**, with their color gradients, provide an intuitive way to see data density, making it easy to identify areas of concentration or sparsity.

These visual tools simplify complex data, turning it into a format that is not only accessible but also actionable.

The Right Tools for the Job

To begin visualizing your data, you need the right tools. Python's Matplotlib and Seaborn libraries are excellent starting points for beginners:

- **Matplotlib** offers basic plotting functions.
- **Seaborn** enhances these with more aesthetically pleasing graphics and additional functionality.

If you prefer a more interactive style, consider **Tableau**. Known for its drag-and-drop capabilities, Tableau allows you to create interactive visualizations **without coding**. It's handy for those who want to focus on data analysis rather than programming.

Power BI is another popular choice, especially if you're familiar with Microsoft products. It integrates data analysis with business intelligence, offering a seamless experience for users looking to dive deep into their data.

Creating effective visualizations involves selecting the right type of chart or graph for your data. **Color** is another powerful tool, enhancing readability and helping viewers quickly grasp key insights.

Scatter plots are ideal for demonstrating relationships between two variables, while **histograms** are perfect for showing data distribution.

When you want to display data density, turn to **heat maps**, which use color to represent different values.

Each type of visualization offers a unique perspective on your data, helping you uncover insights that might not be visible in raw numbers. As you create these visualizations, keep best practices in mind. Choose chart types that match your data to ensure clarity and effectiveness.

Avoid clutter by keeping your visuals simple and focused, allowing the data to speak for itself without distractions.

Multiple-Choice Questions

1. Why is data visualization compared to a detective's magnifying glass?

a. It makes data look more attractive.
b. It helps you discover hidden patterns and insights in datasets.
c. It replaces the need for data analysis.
d. It simplifies coding data processes.

Answer: b. It helps you discover hidden patterns and insights in datasets.

2. What is the primary purpose of translating raw data into visual formats?

a. To make charts look aesthetically pleasing
b. To create complex algorithms for data processing
c. To identify trends, spot anomalies, and make informed decisions easily
d. To store data in a database

Answer: c. To identify trends, spot anomalies, and make informed decisions easily.

3. Which visualization technique is ideal for demonstrating relationships between two variables?

a. Heat map
b. Histogram
c. Pie chart
d. Scatter plot

Answer: d. Scatter plot

4. What type of visualization is best for showing the distribution of a dataset?

a. Histogram
b. Heatmap
c. Line graph
d. Bar chart

Answer: a. Histogram

5. Which library is mentioned as an excellent starting point for data visualization in Python?

a. Seaborn
b. NumPy
c. TensorFlow
d. Pandas

Answer: a. Seaborn

6. What is one of the main advantages of using Tableau for data visualization?

a. It requires extensive coding knowledge.
b. It is primarily a database management tool.
c. It offers drag-and-drop capabilities for creating interactive visualizations.
d. It is only suitable for large datasets.

Answer: c. It offers drag-and-drop capabilities for creating visualizations that are interactive.

7. What role does color play in data visualization?

a. It distracts from the data.
b. It complicates the interpretation of data.
c. It is irrelevant to the quality of the visualization.
d. It enhances readability and helps viewers grasp key insights quickly.

Answer: d. It enhances readability and helps viewers grasp key insights quickly.

8. Which of the following practices should be avoided when creating visualizations?

a. Adding excessive elements that clutter the visualization
b. Choosing chart types that match the data
c. Keeping visuals simple and focused
d. Using color to enhance understanding

Answer: a. Adding excessive elements that clutter the visualization

9. What is the purpose of using a heat map in data visualization?

a. To show data in a linear progression
b. To represent data density using color gradients
c. To compare categorical data side by side
d. To display data correlations in a pie format

Answer: b. To represent data density using color gradients

10. What should you consider when selecting the type of chart or graph for your data?

a. The latest design trends
b. Personal preferences and style
c. The nature of your data and the insights you want to convey
d. The software you are using

Answer: c. The nature of your data and the insights you want to convey

Interactive Visualization Exercise

Take a moment to explore different visualization tools.

Choose a dataset you're interested in and try creating a scatter plot using Matplotlib or Seaborn.

Experiment with various colors and styles to see how they affect the visual's readability.

Once you're comfortable, try building an interactive dashboard in Tableau or Power BI.

As you work through this exercise, reflect on the insights you uncover and consider how these visualizations can inform your decisions or spark new questions about your data.

4.2 CLEANING AND PREPROCESSING DATA FOR ANALYSIS

Imagine setting out to paint a masterpiece, only to realize your canvas is covered in smudges. Working with messy data is much the same; it obscures the true picture, leading to inaccurate analysis and unreliable models.

Clean data is the bedrock of effective data analysis, ensuring your models are trained on accurate and consistent information. It's like polishing a lens through which you view the world, sharpening the focus, and clarifying the details.

- **Error detection** is crucial in this process, helping identify inaccuracies that could skew results. By catching these errors early, you can correct them, ensuring the data's integrity before it goes into model training.
- **Consistency checks** further enhance this by ensuring uniformity across data entries and ensuring all pieces align like well-fitted puzzle pieces.

Once you've identified these issues, you can apply common data-cleaning techniques to tidy up your dataset.

- Handling **missing values** is often the first step, as incomplete data can lead to misleading conclusions. Techniques like **imputation**, which fills in gaps with estimated values, or simply removing incomplete entries, help maintain data quality.
- **Data type conversion** is another critical task, ensuring that each data point is in the correct format for analysis. This might involve converting text entries into numerical values, allowing for mathematical operations and comparisons.
- **Outlier detection** is equally important. These are data points that deviate significantly from the rest, potentially indicating errors or unusual occurrences. By identifying and managing these outliers, you protect your analysis from skewed results, ensuring that your model learns from the most accurate representation of your data.

Preprocessing methods go beyond cleaning to enhance the quality and utility of your data.

- **Feature engineering**, for instance, involves creating new variables that can improve a model's predictive power. It's a bit like adding spices to a recipe; introducing new flavors enriches the final dish.
- **Normalization and scaling** adjust numerical data to a standard range, ensuring that no single feature dominates the learning process. This leveling of the field allows your model to consider all variables fairly, enhancing its performance and accuracy.

These preprocessing steps transform raw data into a refined dataset ready for the rigors of analysis, boosting the overall effectiveness of your AI models.

Automation Tools

Automation tools can streamline your approach to data cleaning, saving time and reducing the potential for human error.

DataWrangler provides a user-friendly interface for organizing and cleaning datasets, making it accessible even to those new to data science. It simplifies complex tasks, which allows you to focus on analysis rather than the minutiae of data preparation.

For those comfortable with coding, **Python scripts** offer a powerful way to **automate repetitive cleaning tasks**. By writing custom code, you can program the cleaning process to suit your specific needs, ensuring consistent and efficient results. These tools act as helpful assistants, guiding you through the complexities of data cleaning and empowering you to focus on the insights your data holds.

Multiple-Choice Questions

1. Why is clean data considered the bedrock of effective data analysis?

   a. It makes visuals more appealing.
   b. It ensures models are trained on accurate and consistent information.
   c. It simplifies the coding process.
   d. It reduces the amount of data collected.

Answer: b. It ensures models are trained on accurate and consistent information.

2. What is the primary purpose of error detection in data cleaning?

   a. To identify inaccuracies that could skew results
   b. To remove duplicate entries only
   c. To enhance the aesthetic quality of the data
   d. To automate data collection

Answer: a. To identify inaccuracies that could skew results

3. What technique can be used to handle missing values in a dataset?

   a. Data aggregation
   b. Outlier detection
   c. Data replication
   d. Imputation or removal of incomplete entries

Answer: d. Imputation or removing incomplete entries

4. For what reason is data type conversion important?

a. To ensure data points are in the correct format for analysis
b. To store data in a text format
c. To increase the size of the dataset
d. To simplify the visualization of data

Answer: a. To ensure data points are in the correct format for analysis

5. What are outliers in a dataset?

a. Duplicate entries
b. Data points that significantly deviate from the norm
c. Missing values
d. Standardized data points

Answer: b. Data points that significantly deviate from the norm

6. What does feature engineering involve?

a. Removing unnecessary data points
b. Standardizing data formats
c. Creating new variables to improve model predictive power
d. Converting numerical data into categories

Answer: c. Creating new variables to improve model predictive power

7. What is the purpose of normalization and scaling in data preprocessing?

a. To compress data for easier storage
b. To reduce the number of variables in the dataset
c. To adjust numerical data to a standard range and ensure fair model consideration
d. To eliminate any non-numeric data

Answer: c. To adjust numerical data to a standard range and ensure fair model consideration

8. How do automation tools like DataWrangler assist in the data-cleaning process?

a. They replace the need for data analysis.
b. They require extensive programming knowledge.
c. They eliminate all manual entry tasks.
d. They provide a user-friendly interface for organizing and cleaning datasets.

Answer: d. They provide a user-friendly interface for organizing and cleaning datasets.

9. What advantage do Python scripts offer in data cleaning?

a. They allow for the automation of repetitive cleaning tasks tailored to specific needs.
b. They require no previous knowledge of programming.
c. They provide graphical visualizations of the data.
d. They simplify the data collection process.

Answer: a. They allow for the automation of repetitive cleaning tasks tailored to specific needs.

10. What does a successful data cleaning and preprocessing empower you to do?

a. Focus solely on data storage
b. Concentrate on the insights that the data holds
c. Avoid all future data collection
d. Increase the amount of data collected unnecessarily

Answer: b. Concentrate on the insights that the data holds

With this knowledge, you're well-equipped to move forward, ready to explore the exciting world of AI modeling and beyond. In the next chapter, we'll explore the intricacies of building your first AI model, where you'll see these skills come to life in new and exciting ways.

5

BUILDING YOUR FIRST MODEL

Building your first AI **model** is a lot like standing in front of a blank canvas, ready to bring a new world to life. It's less about memorizing complex theories and more about painting with data and algorithms.

Even if you're new to AI, selecting the right model is your first step toward creating a masterpiece that makes sense of the information around you. It starts with understanding the basics of model selection, where simplicity is key.

As a beginner, you'll want to choose models that are not only easy to understand but also easy to implement. Think of these models as your training wheels, helping you gain confidence as you learn to ride the AI bicycle.

5.1 CHOOSING THE RIGHT AI MODEL FOR BEGINNERS

When selecting a model, you must also **balance accuracy and speed**. Some models are highly accurate but take longer to process data, while others are faster but might sacrifice some precision. It's a bit like choosing between a sports car and a family sedan—each has its strengths, and the right choice depends on your specific needs and circumstances.

Keep in mind the size and type of your data as well. A model compatible with your dataset ensures smoother processing and more reliable results.

For instance, a simple model like linear regression might be ideal if you're working with a small dataset. But if you have a larger, more complex dataset, you might need a model that can handle the additional intricacies.

As a beginner, you'll likely start with models that are both accessible and powerful:

- **Linear regression** is a classic choice for predicting continuous outcomes such as sales forecasts. It's like drawing a straight line through your data points, helping you quickly see trends and make predictions.
- **Decision trees** are another beginner-friendly option. They provide a visual layout to classification tasks, allowing you to make decisions based on a series of if-then rules. Picture a flowchart that guides you through a series of choices, each branch representing a possible decision path.
- **K-means clustering** is an excellent introduction for those interested in grouping data points. This method helps you organize data into clusters based on similarities, like sorting different colored marbles into separate jars.

Evaluating Performance

Once you've chosen a model, evaluating its performance is crucial to ensure its effectiveness before implementation.

Cross-validation is a method that helps you test your model's reliability by splitting your data into multiple subsets. This allows you to train and test the model on each one which gives you a more accurate way to see your model's performance on new, unseen data.

Error metrics like mean squared error and accuracy provide additional insight into your model's performance, helping you identify areas for improvement.

Mean-squared error measures the average squared differences between predicted and actual values, giving you a sense of the model's precision.

To bring these concepts to life, consider some practical examples of where these models excel. Linear regression is often used in **predictive analytics**, such as predicting future sales based on historical data. It's like predicting the weather by analyzing past patterns, allowing businesses to make informed decisions. Decision trees can be applied to **customer segmentation**, helping marketers identify distinct groups within their audience and plan strategies accordingly.

Imagine more personalized marketing efforts by being able to categorize your customers based on their patterns and preferences.

K-means clustering is valuable in **market segmentation**, providing insights into customer groups with similar characteristics. By understanding these clusters, companies can develop campaigns that resonate more effectively with their audience.

Multiple-Choice Questions

1. What is a crucial factor to consider when selecting an AI model?

   a. The color of the model
   b. The operating system used for training
   c. The popularity of the model
   d. The balance between accuracy and speed

Answer: d. The balance between accuracy and speed

2. How is the choice of an AI model compared in the text?

   a. Like choosing a programming language
   b. Like choosing between a sports car and a family sedan
   c. Like selecting a restaurant
   d. Like picking a vacation destination

Answer: b. Like choosing between a sports car and a family sedan

3. Which model is suggested as a suitable option for predicting continuous outcomes?

a. Decision tree
b. K-means clustering
c. Linear regression
d. Support vector machine

Answer: c. Linear regression

4. What type of tasks do decision trees help with?

a. Classification tasks based on if-then rules
b. Analyzing financial data
c. Unsupervised learning
d. Predicting time series data

Answer: a. Classification tasks based on if-then rules

5. What is k-means clustering used for?

a. Predicting future sales
b. Classification of continuous variables
c. Grouping data points into clusters based on similarities
d. Block randomization in experiments

Answer: c. Grouping data points into clusters based on similarities

6. What is the purpose of cross-validation in model evaluation?

a. To increase the size of the dataset
b. To measure the processing speed of the model
c. To create more complex models
d. To test the model's reliability by using multiple subsets of data

Answer: d. To test the model's reliability by using multiple subsets of data

7. Which metric measures the average squared differences between predicted and actual values?

a. Mean-squared error
b. Recall
c. Precision
d. F1 score

Answer: a. Mean-squared error

8. In what scenario is linear regression typically used?

a. Customer segmentation
b. Image recognition
c. Predictive analytics for forecasting sales
d. Natural language processing

Answer: c. Predictive analytics for forecasting sales

9. How can decision trees be applied in marketing?

a. For stock price prediction
b. To identify distinct customer segments and plan strategies
c. To analyze temperature changes over time
d. For processing images in real time

Answer: b. To identify distinct customer segments and plan strategies

10. What benefit does k-means clustering provide in market segmentation?

a. It simplifies product designs.
b. It enhances the website user experience.
c. It reduces the need for data collection.
d. It helps identify distinct customer groups with similar characteristics.

Answer: d. It helps identify distinct customer groups with similar characteristics.

Interactive Exercise: Model Selection Practice

To solidify your understanding, let's try a simple exercise.

Think of a problem you'd like to solve with AI—perhaps predicting student grades or classifying email spam.
Consider the data you might have and the outcome you want to achieve.
Choose a model that fits your needs based on what you've learned.
Describe why you selected this model and how it will help you achieve your goal.

This exercise will help you apply the concepts and gain confidence in your model-selection skills.

5.2 SETTING UP YOUR FIRST AI MODEL: A STEP-BY-STEP GUIDE

Think of setting up your first AI model as preparing to bake a cake. You need the right ingredients and tools as well as a straightforward recipe.

The first step is **environment preparation**, ensuring you've got all the necessary libraries and tools installed on your computer. Think of this as your laboratory or kitchen equipped with everything you need to start creating:

- You'll want a programming language like **Python** ready
- And libraries such as **NumPy** for numerical operations
- **Pandas** for data manipulation
- **Scikit-learn** for machine learning

These libraries are like your flour, sugar, and eggs—a must-have for any model-building recipe.

Once your environment is ready, it's time to integrate your dataset, which involves loading your data and dividing it into separate sets for training and testing. Think of this as dividing your ingredients: some for practice runs and some for the final bake. The training set is where your model learns, and the testing set is where you evaluate its performance.

Ensuring that your model can generalize well to new data is crucial. Most beginners find it helpful to use the **train_test_split** function from **Scikit-learn**, which simplifies this process with just a few lines of code.

Data preprocessing is a key stage before your model can start learning:

- Here, you'll focus on **feature selection**, which involves identifying the most relevant variables to include as inputs for your model. Consider it like choosing which flavors to highlight in your cake. Not every ingredient is necessary, and some might overshadow others.
- You'll also need to transform categorical data into a numerical format, a process known as **data encoding**. This step is vital because most machine learning models work with numerical data. Techniques like **one-hot encoding** can convert categories into **binary vectors**, ensuring your model can process them effectively.

With your data prepped, you move on to **model configuration**, which involves setting up model parameters and tuning hyperparameters to ensure optimal performance:

- **Hyperparameters**, like the temperature and timing for your bake, include settings such as the learning rate or the number of trees in a forest model. Adjusting these parameters can significantly impact how well your model learns from the data.
- Additionally, you'll need to decide on the **model architecture**, which dictates the structure and complexity of your model. The architecture is like your recipe's method—a blueprint guiding how the ingredients combine to create the final product.

Multiple-Choice Questions

1. What is the first step in setting up your AI model?

   a. Environment preparation by installing necessary libraries and tools
   b. Preparing the dataset
   c. Configuring model parameters
   d. Evaluating model performance

Answer: a. Environment preparation by installing necessary libraries and tools

2. Which programming language is suggested for building AI models?

   a. Java
   b. C++
   c. Ruby
   d. Python

Answer: d. Python

3. What is the purpose of the train_test_split function in Scikit-learn?

   a. To divide data into training and testing sets
   b. To evaluate model performance
   c. To load datasets from the internet
   d. To perform data visualization

Answer: a. To divide data into training and testing sets

4. Why is data preprocessing important before model training?

   a. It helps visualize the data.
   b. It ensures the model can generalize well to new data.
   c. It simplifies the coding process.
   d. It reduces data size.

Answer: b. It ensures the model can generalize well to new data.

5. What does feature selection involve?

a. Choosing all available data points for the model
b. Ensuring data is formatted correctly
c. Randomly selecting features
d. Identifying the most relevant variables for model inputs

Answer: d. Identifying the most relevant variables for model inputs

6. What is one method for transforming categorical data into a numerical format?

a. Normalization
b. Imputation
c. One-hot encoding
d. Min-max scaling

Answer: c. One-hot encoding

7. What are hyperparameters in the context of AI model configuration?

a. Settings that are adjusted during model training
b. Fixed parameters that define the model's performance
c. Variables related to dataset size
d. Constant values used only for visualization

Answer: a. Settings that are adjusted during model training

8. What does "model architecture" refer to?

a. The coding language used for implementation
b. The structure and complexity of the model
c. The amount of training data used
d. The visualization tools employed

Answer: b. The structure and complexity of the model

9. Why is it important to tune hyperparameters?

a. It ensures the model can process textual data.
b. It can significantly impact the model's learning performance.
c. It eliminates the need for preprocessing.
d. It guarantees perfect accuracy in every case.

Answer: b. It can significantly impact the model's learning performance.

10. In the cake-baking analogy, what are hyperparameters compared to?

a. The type of cake being made
b. The ingredients used
c. The temperature and timing for baking
d. The decoration of the cake

Answer: c. The temperature and timing for baking

Interactive Exercise: Decision Tree Model Practice

To illustrate, let's walk through a practical example of building a decision tree model for classifying email spam:

First, you'll gather a dataset containing emails labeled as "spam" or "not spam."

After preparing your environment and loading the data, you'll split it into training and testing sets.

Next, you'll preprocess the text data using tokenization and TF-IDF encoding techniques, which convert text into a numerical format the model can understand.

With preprocessing complete, you'll configure your decision tree model, selecting hyperparameters like maximum depth to prevent overfitting.

As you build and train the model, you'll evaluate its performance using your testing set, making adjustments as necessary to improve accuracy.

This hands-on technique reinforces the setup process and provides a concrete understanding of building and refining models. The experience you gain here will serve as a foundation for more complex models and prepare you for future AI projects involving advanced techniques or larger datasets.

5.3 TRAINING BASICS: HOW TO TRAIN YOUR FIRST MODEL

Have you ever trained a dog to fetch a ball? Initially, there's some trial and error, but eventually, your dog learns to associate the command with the action. Similarly, training an AI model involves teaching it to recognize patterns through exposure to data.

This process is crucial in AI because it transforms raw data into a model capable of making predictions or decisions. Training is about repetition and reinforcement, allowing the model to identify relationships within the data. Each time you run the training process, the model refines its understanding, just like your dog gets better at fetching with practice.

When it comes to training methodologies, you have a couple of paths to explore, each with its own approach and relevance to newcomers:

- **Supervised learning** is like giving your model a guided tour with labeled data. It involves feeding the model data that includes both inputs and the correct outputs, much like showing students the answers to problems before asking them to solve similar ones. This method is incredibly useful for tasks where you have a clear idea of what the output should be, such as classifying emails or predicting house prices.
- On the other hand, **unsupervised learning** is more of an exploratory journey. Here, you provide the model with data that lacks clear labels, allowing it to find patterns and groupings independently. This is like letting a detective loose in a mystery, piecing together clues without knowing the whole story. This method is ideal for uncovering hidden structures in data, like market segmentation in customer data.

Once your model begins training, evaluating its performance becomes critical. Here, you assess how well the model is learning and whether adjustments are needed. During training, you split your data into training and validation sets.

The **training set** is where the model learns, while the **validation set** tests its generalization ability. It's like studying for an exam and then taking a mock test to see how well you understand the material. This split helps prevent **overfitting**, where the model performs well on training data but not so well with new data.

Another key concept is the **loss function**, which acts like a report card for your model. It measures how far off the model's predictions are from the actual outcomes, providing you with feedback for improvement. A lower loss indicates a better-performing model, guiding you to fine-tune its learning process.

Training Challenges

Training isn't without its challenges, but with a few tips, you can navigate them successfully. One common hurdle is overfitting, where the model learns the training data too well and does not learn to generalize.

To combat this, you can use techniques like **regularization**, which adds a penalty for complexity, encouraging the model to stay simple.

Another challenge is **underfitting**, where the model is too simplistic and misses essential patterns. In this case, you should increase the model's complexity or provide more data for training. Monitoring the **learning rate**, which controls how much the model adjusts its weights during training, is also crucial. A rate that's too high can cause the model to overshoot optimal solutions, while a rate that's too low can slow down learning. Keeping a watchful eye on these factors ensures your model trains effectively, paving the way for successful AI applications.

Multiple-Choice Questions

1. What is the primary goal of training an AI model?

   a. To create visually appealing data visualizations
   b. To collect more data
   c. To develop complex algorithms
   d. To teach the model to recognize patterns through exposure to data

Answer: d. To teach the model to recognize patterns through exposure to data

2. How is supervised learning best described?

a. Providing the model with labeled data to learn from
b. An exploratory journey without clear guidance
c. A method solely used for classification tasks
d. Training without any prior information

Answer: a. Providing the model with labeled data to learn from

3. What analogy is used to describe unsupervised learning?

a. Teaching a class with set answers to problems
b. Following a recipe step-by-step
c. Running a marathon without training
d. Letting a detective piece together clues independently

Answer: d. Letting a detective piece together clues independently

4. Why is it important to split your data into training and validation sets?

a. To increase the amount of data available
b. To see how well the model generalizes to new data and prevents overfitting
c. To reduce the complexity of data
d. To enhance the visual representation of data

Answer: b. To see how well the model generalizes to new data and prevents overfitting

5. What does the loss function measure during model training?

a. The complexity of the model
b. The success of the data collection process
c. How far off the model's predictions are from actual outcomes
d. The generalization ability of the model

Answer: c. How far off the model's predictions are from actual outcomes

6. What is overfitting in the context of training an AI model?

a. When the model learns too little from the training data
b. When the model learns the training data too well but fails to generalize to new data
c. When the model is unable to learn from the given data
d. When the training data is too simplistic

Answer: b. When the model learns the training data too well but fails to generalize to new data

7. What technique can be used to combat overfitting?

a. Decreasing data complexity
b. Increasing the size of the training set
c. Regularization, which adds a penalty for complexity
d. Using fewer features

Answer: c. Regularization, which adds a penalty for complexity

8. Which problem occurs when a model is too simplistic and misses essential patterns?

a. Overfitting
b. Underfitting
c. Validation error
d. Loss of data

Answer: b. Underfitting

9. What role does the learning rate play in model training?

a. It determines how quickly a model can learn from data.
b. It decides the model architecture.
c. It measures model accuracy.
d. It adjusts the amount of data collected.

Answer: a. It determines how quickly a model can learn from data.

10. What should you do if the learning rate is set too high?

a. Increase the amount of training data.
b. Decrease the model complexity.
c. Watch for overshooting optimal solutions.
d. Modify the loss function.

Answer: c. Watch for overshooting optimal solutions.

As you wrap up this chapter, remember that learning AI is a journey of exploration and growth. While you're learning about AI, you're also becoming part of a community of thinkers and creators, shaping the future of technology.

Now that you've built and trained your model, it's time to explore how AI models evolve and improve, stepping into a world where continuous learning and adaptation are key to success.

6

EVALUATING AND IMPROVING MODELS

Picture this: You've just baked a cake following a new recipe. The aroma is delightful, but how do you know it's genuinely delicious? You take a bite and assess the texture, sweetness, and overall taste.

In AI, evaluating a model's performance is much like that first taste test. It's where you determine whether your AI model delivers on its promises or needs more tweaking. It is essential to evaluate the performance of your model to ensure that it is functioning and achieving its intended purpose accurately and effectively.

6.1 MEASURING SUCCESS: EVALUATING MODEL PERFORMANCE

One primary tool in your evaluation toolkit is **metrics**. Consider them the criteria judges use in a talent show to rate performances. Metrics like accuracy, precision, and recall help you quantify how well your model is doing:

- **Accuracy** measures the percentage of correct predictions from the total number of cases.
- **Precision** focuses on the quality of positive predictions—how many of the predicted positives were actual positives.

- **Recall** is a measurement of a model's ability to identify all relevant instances. Balancing these metrics is crucial, especially in classification tasks, where the stakes can be high.
- The **F1 score** is used to provide a comprehensive evaluation. It's a harmonic means of precision and recall, offering a single metric that balances both. It becomes especially useful when precision and recall need to be balanced, ensuring your model performs well across multiple dimensions.

Alongside metrics, **evaluation techniques** play a vital role in validating model effectiveness. The **confusion matrix** is a powerful visualization tool that breaks down predictions into true positives, true negatives, false positives, and false negatives.

Imagine a grid that helps you see where your model hits and misses, like a sports coach analyzing a game performance. This matrix allows you to pinpoint areas for improvement and adjust your strategy accordingly.

Another essential tool is the **receiver operating characteristic (ROC) curve**, which represents a model's performance across various thresholds. It's like adjusting the volume on a stereo to find the perfect sound level. By plotting true positive rates against false positive rates, the ROC curve helps you visualize and select optimal thresholds, enhancing decision-making.

Cross-validation strategies further boost model evaluation by ensuring hardiness. **K-fold cross-validation** involves dividing your dataset into **k subsets**, training the model on **k-1 subsets**, and testing the remaining one. Think of it as a dress rehearsal, using all the data for training and testing.

Leave-one-out cross-validation takes this a step further, using a single observation for testing while using the rest for training, providing a more granular evaluation.

In practical scenarios, these evaluation techniques prove invaluable. Consider a spam filter: Precision and recall are critical to balancing false alarms and missed spam. **High precision** ensures that flagged emails are indeed spam, while **high recall** captures all spam emails. Similarly, metrics can improve user satisfaction in a recommendation system by ensuring that suggested items are relevant and engaging, like a well-curated playlist that hits all the right notes.

Multiple-Choice Questions

1. Why is evaluating an AI model's performance compared to tasting a cake?

a. To describe the cooking process
b. To ensure the model delivers on its promises and aims
c. To measure the ingredients used
d. To simplify the training process

Answer: b. To ensure the model delivers on its promises and aims

2. What do model evaluation metrics help quantify?

a. The number of features in the model
b. The amount of training data
c. How well the model is functioning
d. The complexity of the model

Answer: c. How well the model is functioning

3. What does accuracy measure in model performance?

a. The rate of false positives
b. The complexity of the data
c. The model's training time
d. The percentage of correct predictions out of total predictions

Answer: d. The percentage of correct predictions out of total predictions

4. How is precision defined in the context of model performance?

a. The quality of positive predictions—how many of the predicted positives were actual positives
b. The overall correctness of predictions
c. The ability to capture all relevant instances
d. The percentage of training data used

Answer: a. The quality of positive predictions—how many of the predicted positives were actual positives

5. What does recall measure in model evaluation?

a. The ratio of correct predictions to total predictions
b. The ability of the model to identify all relevant instances
c. The complexity of the model
d. The model's response time

Answer: b. The ability of the model to identify all relevant instances

6. What is the purpose of the F1 score?

a. To measure training speed
b. To balance precision and recall into a single metric
c. To determine the number of features in the model
d. To visualize performance across different datasets

Answer: b. To balance precision and recall into a single metric

7. How does a confusion matrix assist in model evaluation?

a. By visualizing the model's architecture
b. By tracking the model's training history
c. By breaking down predictions into true positives, true negatives, false positives, and false negatives
d. By providing dynamic data visualization options

Answer: c. By breaking down predictions into true positives, true negatives, false positives, and false negatives

8. What does the ROC curve represent?

a. The time taken for model training
b. The precision and recall of the model simultaneously
c. The overall accuracy of the model
d. A graphical representation of true positive rates against false positive rates across various thresholds

Answer: d. A graphical representation of true positive rates against false positive rates across various thresholds

9. What is K-fold cross-validation?

a. A method of partitioning the dataset into K subsets for thorough evaluation
b. A technique that uses all available data for training
c. A process to measure the complexity of the model
d. An approach to reduce false negatives

Answer: a. A method of partitioning the dataset into K subsets for thorough evaluation

10. In the context of a spam filter, why are precision and recall particularly important?

a. High precision and high recall balance false alarms and missed spam.
b. They determine the speed of the model.
c. They measure user engagement.
d. They help visualize data patterns.

Answer: a. High precision and high recall balance false alarms and missed spam.

Reflection Section: Evaluating Your AI Model

Take some time to reflect on a model you're working with or are interested in.

What metrics would be most relevant for its evaluation?
Consider accuracy, precision, recall, and the F1 score.
How might a confusion matrix or ROC curve help you visualize its performance?

As you think through these questions, jot down your thoughts.
This reflection can guide your understanding and application of model evaluation techniques, ensuring you can effectively assess and improve your AI models.

6.2 ITERATIVE LEARNING: HOW TO IMPROVE YOUR MODEL

Let's say you're learning to play a musical instrument. Every time you practice, you listen closely to the sound you produce, making small adjustments to improve. This same concept applies to iterative learning in AI, where continuous learning and adaptation can significantly enhance your model's performance.

The feedback loop is at the heart of this process, where model outputs are analyzed to refine inputs and strategies. You can tweak its parameters by understanding where the model falls short, much like tuning a guitar to achieve the perfect pitch.

Another crucial aspect of iterative learning is adjusting the learning rate, which determines how quickly or slowly a model updates its understanding. If the learning rate is too high, the model might overshoot optimal solutions, making it erratic. On the other hand, a too-low rate can slow down learning, causing the model to get stuck in its ways.

It's a balancing act, like adjusting the speed of a metronome to keep the rhythm just right. By fine-tuning the learning rate, you ensure that your model learns efficiently and effectively, adapting to new data with agility.

Hyperparameter tuning is another key player in model optimization. Think of hyperparameters as the settings on a car's dashboard, where each one influences how the vehicle performs. Adjusting these settings can create significant improvements in model accuracy and efficiency. Two popular methods for tuning hyperparameters are grid search and random search:

- **Grid search** involves an exhaustive search over specified parameter values, offering a systematic approach to finding the best combination.
- **Random search** takes a different route, exploring random combinations to identify effective configurations. This method can be most useful when dealing with large datasets, offering quicker insights into optimal settings.

Feature selection and engineering also play pivotal roles in model improvement. Identifying and selecting impactful features can enhance model accuracy, as well:

- Techniques such as **feature importance ranking** help you pinpoint which variables truly matter, allowing you to focus on those that add the most value.
- **Creating interaction terms** is another powerful technique, combining existing features to capture more complex relationships within the data.

To illustrate the power of iterative learning, consider a predictive maintenance model designed to forecast equipment failures. Initially, the model might struggle with accuracy, but through **feature engineering**—such as incorporating sensor data and environmental factors—predictions can improve dramatically.

Similarly, take a sentiment analysis tool that relies on user feedback to refine its understanding of language nuances. By iterating on the model, incorporating feedback, and adjusting features, the tool becomes increasingly adept at capturing the subtleties of human emotion.

Iterative learning is a dynamic and rewarding process that offers endless opportunities for model enhancement. It's about embracing change and being open to experimentation, knowing that each iteration brings you closer to an optimal solution. With patience and persistence, you can transform your AI models from basic prototypes into sophisticated systems capable of addressing complex challenges.

Multiple-Choice Questions

1. What analogy is used to explain the concept of iterative learning in AI?

   a. Learning to cook
   b. Learning to play a musical instrument
   c. Building a house
   d. Writing a book

Answer: b. Learning to play a musical instrument

2. What is the core component of the iterative learning process?

   a. Feedback loop
   b. Data collection
   c. Initial model training
   d. Feature selection

Answer: a. Feedback loop

3. What happens if the learning rate is set too high?

   a. The model learns too slowly.
   b. The model improves accuracy significantly.
   c. The model gets stuck in its local minimum.
   d. The model becomes erratic and overshoots optimal solutions.

Answer: d. The model becomes erratic and overshoots optimal solutions.

4. What is hyperparameter tuning compared to?

   a. The process of feature selection
   b. The cooking temperature in a recipe

c. The settings on a car's dashboard
d. The style of written communication

Answer: c. The settings on a car's dashboard

5. Which of the following are two popular methods for tuning hyperparameters?

a. Manual adjustment and automated testing
b. Grid search and random search
c. Linear regression and logistic regression
d. Feature selection and data cleaning

Answer: b. Grid search and random search

6. How does feature selection enhance model accuracy?

a. By identifying and selecting impactful features
b. By removing all variables from the dataset
c. By transforming numerical data into categorical data
d. By duplicating existing features

Answer: a. By identifying and selecting impactful features

7. What technique involves combining existing features to capture complex relationships?

a. Data normalization
b. Data aggregation
c. Dimensionality reduction
d. Feature interaction

Answer: d. Feature interaction

8. Which example illustrates the application of iterative learning in predictive maintenance?

a. Analyzing user demographics for marketing
b. Simplifying the model to reduce computation
c. Incorporating sensor data and environmental factors to improve predictions
d. Collecting historical failure data only

Answer: c. Incorporating sensor data and environmental factors to improve predictions

9. What is the overall attitude encouraged toward the iterative learning process?

a. Dismiss experimentation and stick to original plans.
b. Embrace change and be open to experimentation.
c. Avoid adjusting models frequently.
d. Rely solely on initial data for decisions.

Answer: b. Embrace change and be open to experimentation.

10. What is the key takeaway about transforming AI models through iterative learning?

a. Each iteration brings you closer to the best solution.
b. It is a quick and easy process.
c. Iterative learning is only necessary for large projects.
d. Once the model is created, no further adjustments are needed.

Answer: a. Each iteration brings you closer to an optimal solution.

6.3 ADDRESSING OVERFITTING AND UNDERFITTING IN MODELS

Imagine fitting a rubber band snugly around a complex, jagged rock. If the band conforms too tightly to every nook and cranny, it might snap at the slightest movement. This is similar to overfitting in models:

- **Overfitting** occurs when a model becomes over trained in the data, capturing noise and minor fluctuations rather than the underlying pattern. This overzealous fitting means that while it performs excellently on training data, its predictions on new, unseen data fall flat. Picture a student who memorizes every word in a textbook but struggles with questions that require critical thinking. The model has essentially memorized rather than generalized.
- Conversely, **underfitting** is like using a straight ruler to measure the curves of the same rock. The ruler fails to capture the intricacies, offering a poor representation. Underfitting is the result of a model that is too simplistic and unable to capture the deeper trends in the data. It's like a student skimming the textbook and missing key concepts, resulting in poor performance.

Both issues—overfitting and underfitting—are common modeling challenges and require careful handling:

- **Regularization** methods can combat overfitting, such as adding guardrails to prevent a car from veering off course.
- **L1 and L2 regularization** introduce a penalty for significant coefficients, encouraging the model to maintain simplicity. It's like gently reminding the model to focus on the broader picture rather than the minutiae.
- Another effective strategy is **pruning** decision trees. Cutting back complex branches prevents the model from becoming too detailed and specific. Pruning simplifies the tree, promoting better generalization.

Conversely, addressing underfitting involves enhancing model complexity, allowing it to capture more intricate patterns. Increasing model layers in neural networks adds depth and complexity, enabling the model to learn rich, complex features.

Similarly, **polynomial regression** introduces polynomial terms to capture **nonlinear relationships** in data. The model can better represent nuanced patterns by fitting curves rather than straight lines.

Consider real-world applications where balancing these issues is crucial. In healthcare, an overfitted model predicting patient outcomes might perform well in a controlled environment but not so well in real-world scenarios where patient conditions vary widely.

Techniques like **regularization** ensure that the model remains adaptable and provides reliable predictions. On the other hand, a financial model used for predicting stock prices might underfit if it fails to account for complex market dynamics. Increasing model complexity through additional layers or features can help capture these intricacies, improving accuracy.

In practice, finding the balance between overfitting and underfitting is an art. It requires experimentation, patience, and a solid understanding of the data and domain. By employing techniques like regularization, pruning, and enhancing model complexity, you can guide your models to achieve the sweet spot: a robust, flexible model that performs well across diverse datasets. This balancing act ensures your models are technically sound and practically valuable, ready to tackle real-world challenges confidently.

Multiple-Choice Questions

1. What analogy is used to explain overfitting?

   a. Using a straightedge to measure curves
   b. Taking a test without studying
   c. Cooking a dish with too many spices
   d. Snapping a rubber band around a jagged rock

Answer: d. Snapping a rubber band around a jagged rock

2. What is overfitting in a model?

   a. Learning too little from the training data
   b. Memorizing the training data too well, including noise
   c. Understanding general patterns in the data
   d. Effectively handling unseen data

Answer: b. Memorizing the training data too well, including noise

3. What is the key characteristic of underfitting?

a. The model predicts new data exceptionally well.
b. The model captures all intricate details of the training data.
c. The model is too simplistic and misses key trends.
d. The model generates complex outputs.

Answer: c. The model is too simplistic and misses key trends.

4. What regularization methods are commonly used to combat overfitting?

a. L1 and L2 regularization
b. K-means clustering
c. Logistic regression
d. Dimensionality reduction

Answer: a. L1 and L2 regularization

5. What is the importance of pruning in decision trees?

a. To make the model more complex
b. To simplify the model and prevent overfitting
c. To increase the number of branches in the tree
d. To enhance the visual representation of the tree

Answer: b. To simplify the model and prevent overfitting

6. How can you address underfitting?

a. By decreasing model complexity
b. By simplifying the dataset
c. By increasing model complexity or adding layers
d. By reducing the amount of training data

Answer: c. By increasing model complexity or adding layers

7. What technique can be used to capture nonlinear relationships in data?

a. Linear regression
b. Data normalization
c. Regularization
d. Polynomial regression

Answer: d. Polynomial regression

8. Why are balancing overfitting and underfitting important in healthcare models?

a. To ensure accurate predictions that apply to real-world scenarios
b. To quickly deploy models without testing
c. To simplify the training dataset
d. To enhance user experience in medical software

Answer: a. To ensure accurate predictions that apply to real-world scenarios

9. What does achieving the "sweet spot" between overfitting and underfitting involve?

a. Confusing the data with irrelevant features
b. Employing techniques without understanding the domain
c. Relying solely on automated processes
d. Experimentation, patience, and understanding of data

Answer: d. Experimentation, patience, and understanding of data

10. What is the overall goal when addressing overfitting and underfitting?

a. To develop a solid, flexible model that performs well across diverse datasets
b. To create a complex model that only works on training data
c. To minimize the amount of data used
d. To standardize all datasets for testing

Answer: a. To develop a solid, flexible model that performs well across diverse datasets

With the completion of this chapter, you have reinforced your understanding of model evaluation and improvement. As you turn the page to the next chapter, you'll dive deeper into the exciting world of AI applications, ready to apply what you've learned to real-world scenarios and challenges.

"The best way to find yourself is to lose yourself in the service of others."

Mahatma Gandhi

People who give without expecting anything in return often find joy and happiness in life. Together, let's make a difference!

Would you help someone who is just like you—curious about learning and exploring AI but feeling a bit lost about where to begin?

My goal is to make learning AI enjoyable for beginners and super simple for everyone.

But to reach more curious minds, I need your support.

You see, many people pick books based on the comments and reviews they find. So, I'm reaching out to you to help a fellow explorer interested in AI by leaving a review.

It doesn't cost a thing and will take less than a minute, but it could have a big impact on someone else's journey into AI. Your review could help...

...one more small business owner provide for their community.
...one more dreamer follow their entrepreneurial path.
...one more worker find fulfilling and meaningful work.
...one more person change their life for the better.
...one more wish come true.

To make a difference, just scan the QR code below and share your thoughts:

[https://www.amazon.com/review/review-your-purchases/?asin=
BOOKASIN]

If you love helping others, you're the kind of person I admire. Thank you from the bottom of my heart for your support!

Eli Coderight

7

MAKING PREDICTIONS AND APPLICATIONS

Imagine you're a weather forecaster tasked with predicting tomorrow's weather. You can access data from satellites, temperature sensors, and historical patterns. Your job is to translate all this information into a simple prediction: sunny, cloudy, or rainy.

This is similar to how AI models make predictions, though on a much broader scale. They take input data, analyze it, and produce predictions that guide decisions across various fields. In AI, **prediction models** work behind the scenes, using algorithms to process data and forecast outcomes, making them invaluable tools in our increasingly data-driven world.

7.1 FROM DATA TO DECISION: MAKING PREDICTIONS WITH AI

The prediction process begins with algorithms, which act as the brains behind the operation. These algorithms range from **simple linear regression**, which predicts outcomes based on linear relationships, to **complex neural networks** that can process large amounts of data to identify intricate patterns.

Each algorithm serves a specific purpose, such as forecasting future trends or classifying data. For example, **decision trees** can be used to classify customer preferences, while **support vector machines** might help identify patterns in financial transactions.

When data flows through these algorithms, the models learn to make predictions based on patterns and relationships, much like how our brains process information.

One fascinating aspect of AI predictions is their ability to operate in real time, offering live insights as new data becomes available. Imagine a store manager who can adjust inventory levels on the fly based on real-time sales predictions.

This capability is especially valuable in dynamic environments like retail, where demand can fluctuate unpredictably. Systems like **Vertex AI Forecast** exemplify this by processing vast datasets and predicting demand with remarkable accuracy. They provide businesses with live insights that help them optimize inventory, manage supply chains, and enhance customer satisfaction, all by anticipating needs before they arise.

Similarly, **predictive maintenance** in manufacturing uses AI to anticipate equipment failures before they occur. By analyzing machine data and performance metrics, AI models can predict when it is likely to fail, allowing for timely maintenance and reducing downtime. This proactive approach not only saves costs but also extends the lifespan of machines, promoting operational efficiency.

Predictions absolutely must be accurate, as even minor errors can have significant consequences.

Confidence intervals play a crucial role here, providing a range within which the true value of a prediction is likely to fall. They offer a level of certainty, helping decision-makers understand the reliability of predictions and make informed choices.

Continuous model updates further enhance accuracy by adapting to new data and changing conditions. These updates make sure that the model remains relevant and accurate over time, much like updating software to improve performance and security.

- You'll need to start with data input preparation to make predictions using AI models. This involves formatting your data to ensure compatibility with the chosen algorithm.
- Clean, structured data is key to accurate predictions, much like using fresh ingredients for a recipe.
- Once the data is ready, the model processes it, producing prediction outputs that reveal insights about future trends or classifications.
- Interpreting these outputs is essential, as they inform the decisions you need to make.
- Understanding the implications of predictions allows you to act confidently, whether it's adjusting inventory levels, scheduling equipment maintenance, or planning marketing strategies.

Multiple-Choice Questions

1. What is the primary role of algorithms in AI prediction models?

a. To format the input data
b. To analyze data and produce predictions
c. To collect data from satellites
d. To interpret prediction outputs

Answer: b. To analyze data and produce predictions

2. Which algorithm is likely to be used for classifying customer preferences?

a. Linear regression
b. Neural networks
c. Decision trees
d. Support vector machines

Answer: c. Decision trees

3. In what way can AI predictions benefit retail businesses?

a. By providing historical data analysis
b. By predicting future employee performance

c. By accurately forecasting product demand
d. By creating marketing content

Answer: c. By accurately forecasting product demand

4. What concept helps decision-makers understand the reliability of predictions?

a. Confidence intervals
b. Continuous model updates
c. Data input preparation
d. Algorithm selection

Answer: a. Confidence intervals

5. What is essential for ensuring accurate predictions in AI models?

a. Using diverse datasets
b. Continuous monitoring of output
c. Implementing hardware upgrades
d. Formatting clean, structured data

Answer: d. Formatting clean, structured data

6. How do real-time AI predictions improve inventory management for a store manager?

a. By reducing the number of items in stock
b. By providing customer feedback
c. By allowing adjustments based on live sales predictions
d. By recording historical sales trends

Answer: c. By allowing adjustments based on live sales predictions

7. What is the purpose of updating AI models continuously?

a. To change the algorithm employed
b. To enhance the model's accuracy over time
c. To simplify the data entry process

d. To eliminate the need for data preparation

Answer: b. To enhance the model's accuracy over time

8. In the context of AI predictions, what is a primary benefit of predictive maintenance in manufacturing?

a. It predicts equipment failures to allow timely maintenance.
b. It keeps costs low by reducing employee hours.
c. It increases production speed by automating processes.
d. It eliminates all machine downtimes.

Answer: a. It predicts equipment failures to allow timely maintenance.

9. Which of the following describes how machine learning models learn to make predictions?

a. By accessing live data streams directly
b. By following static programming rules
c. By storing historical data without analysis
d. By analyzing large datasets for patterns and relationships

Answer: d. By analyzing large datasets for patterns and relationships

10. Why is data input preparation critical before using an AI model?

a. It ensures compatibility with the chosen algorithm.
b. It reduces the time taken to analyze data.
c. It guarantees that predictions are always accurate.
d. It eliminates the need for a confidence interval.

Answer: a. It ensures compatibility with the chosen algorithm.

Interactive Exercise: Prediction Practice

Consider a simple prediction you might want to make, such as forecasting next week's weather or predicting the demand for a new product

Choose a dataset that fits your scenario
Identify an algorithm that can help make your prediction.
Consider the variables that might influence your prediction and how you'll interpret the results.

This exercise will help you apply what you've learned and build confidence in your prediction-making skills.

7.2 APPLYING AI MODELS IN EVERYDAY SCENARIOS

Who doesn't love the idea of starting their day with a perfectly orchestrated schedule, their smart calendar rearranging meetings to fit in a surprise lunch invitation from a friend? AI tools have revolutionized **personal productivity**, making this kind of seamless day-to-day planning possible.

Let's consider practical examples of AI's impact.

- **Smart-scheduling applications** analyze your habits and preferences to optimize your daily workflow. By understanding your routine, these AI-driven tools can suggest the best times for meetings, remind you of deadlines, and even propose breaks to enhance productivity. This technology minimizes decision fatigue, allowing you to focus on what truly matters.
- Meanwhile, in **entertainment and shopping**, AI-driven personalized recommendations suggest content and products customized to your tastes. Whether it's a new series to binge-watch or a pair of shoes that match your style, these suggestions enhance your experience by ensuring you spend less time searching and more time enjoying.

- AI models are also making waves in enhancing user experiences across other domains. Take **ecommerce**, for instance, where **personalization** has become the norm. AI analyzes your browsing history, purchase behavior, and even your social media activity to create shopping experiences that feel unique to you. This not only boosts customer satisfaction but also increases sales for businesses.
- Similarly, chatbots have become invaluable in **customer service**. These AI-powered virtual assistants provide instant support, answering questions and resolving issues without human intervention. They're available 24/7, offering a level of convenience that traditional customer service methods can't match.
- Where **health and wellness** are concerned, AI models are transforming how we monitor and improve our personal health. **Fitness-tracking devices** with AI capabilities analyze your sleeping patterns, activity levels, and even your heart rate to provide personalized fitness plans. These insights help you set realistic goals and track your progress, ensuring you stay motivated.
- **Health-monitoring apps** take it further by using predictive analytics to identify potential health issues early on. By analyzing data from wearables and other health devices, these apps can alert you to potential risks, allowing you to take preventive measures. With this technology, you can take control of your health, leading to better outcomes and a healthier lifestyle.
- In **travel planning**, AI systems analyze your preferences, past trips, and current trends to suggest itineraries that cater to your unique interests. Whether you're an adventure seeker or a culture enthusiast, these automated suggestions ensure your travel plans are both exciting and hassle-free.
- **In-home automation** and smart systems learn from your habits and adjust settings to match your preferences. From setting the thermostat to dimming the lights, these systems create a comfortable and energy-efficient living environment without you lifting a finger. By adapting to your lifestyle, AI-powered home

automation enhances convenience and promotes sustainability by optimizing energy use.

AI's integration into our daily routines show its potential to improve our quality of life. It's making tasks easier, interactions smoother, and experiences more personalized. As these technologies continue to evolve, the possibilities for enhancing our everyday scenarios are virtually limitless, promising a future where AI is part of our daily lives.

Multiple-Choice Questions

1. How do smart-scheduling applications improve personal productivity?

a. By suggesting optimal times for activities based on habits and preferences
b. By eliminating all meetings
c. By reminding users of events they forgot
d. By blocking time for personal activities only

Answer: a. By suggesting optimal times for activities based on habits and preferences

2. What is the primary benefit of AI-driven personalized recommendations in entertainment?

a. They increase the cost of subscriptions.
b. They help users find content that matches their tastes quickly.
c. They provide a broader range of options without personalization.
d. They limit choices to popular items only.

Answer: b. They help users find content that matches their tastes quickly.

3. In which area have AI chatbots become particularly valuable?

a. Cooking and meal preparation
b. Academic tutoring
c. Athletic Coaching
d. Customer service and support

Answer: d. Customer service and support

4. How do fitness-tracking devices utilize AI capabilities?

a. They set random fitness goals for users.
b. They analyze user data to provide personalized fitness plans.
c. They create generic exercise regimens for everyone.
d. They monitor only step count.

Answer: b. They analyze user data to provide personalized fitness plans.

5. What advantage does AI have in health-monitoring applications?

a. It provides only general health tips without data analysis.
b. It offers personalized health insights based solely on diet.
c. It uses predictive analytics to identify potential health issues early.
d. It replaces traditional medical advice completely.

Answer: c. It uses predictive analytics to identify potential health issues early.

6. What role does AI play in travel planning?

a. It removes the need for any manual input when booking.
b. It guarantees the lowest prices for all travel arrangements.
c. It suggests only the most popular tourist destinations.
d. It generates travel plans that are catered to individual preferences and trends.

Answer: d. It generates travel plans that are catered to individual preferences and trends.

7. How can AI contribute to home automation?

a. By requiring manual overrides for every adjustment
b. By limiting the functionality of smart devices
c. By learning user habits to adjust settings automatically
d. By creating cumbersome settings that confuse users

Answer: c. By learning user habits to adjust settings automatically

8. What is the potential impact of AI integration into daily routines?

a. It makes tasks more complex and time-consuming.
b. It improves the quality of life by making tasks easier and experiences more personalized.
c. It reduces the need for technology altogether.
d. It replaces human interaction in all scenarios.

Answer: b. It improves the quality of life by making tasks easier and experiences more personalized.

9. Why is minimizing decision fatigue important in personal productivity?

a. Reducing choices allows individuals to focus on what truly matters.
b. It prevents people from making any decisions.
c. It promotes independent decision-making.
d. It increases the number of decisions that must be made daily.

Answer: a. Reducing choices allows individuals to focus on what truly matters.

10. What is one notable feature of AI-driven ecommerce personalization?

a. It solely relies on user reviews for recommendations.
b. It stops recommending products after a purchase.
c. It ensures all customers see the same product suggestions.
d. It analyzes browsing history and purchase behavior personalized to experiences.

Answer: d. It analyzes browsing history and purchase behavior personalized to experiences.

7.3 CREATIVE APPLICATIONS: USING AI IN NOVEL WAYS

We now live in a world where machines can create art, compose symphonies, and even write poetry. AI has begun to blur the lines between human creativity and computational power, offering tools that push the limits of traditional art and design.

- **Artists** now use AI to generate unique pieces, exploring styles and techniques that might not be possible with human hands alone. These AI tools assist in creating visually stunning artwork by analyzing vast databases of images and learning patterns and styles.
- In **music**, AI-assisted composition tools can generate melodies and rhythms, providing musicians with fresh ideas and new creative directions. It's like having a collaborator who never tires and is always ready to offer inspiration at the click of a button. These innovations don't replace human creativity; they enhance it, offering new possibilities and pushing artists and musicians to explore uncharted territories.
- In **scientific research**, AI is rapidly becoming an indispensable tool. Consider **drug discovery**, where AI models accelerate pharmaceutical research by predicting how different compounds might interact with biological targets. This predictive power enables researchers to focus on the most promising candidates, accelerating the development of new medications and potentially saving lives.
- AI's role extends beyond the lab, impacting **environmental monitoring** as well. By analyzing satellite data and sensor readings, AI models precisely track climate-change impacts, offering insights into **temperature shifts**, **deforestation**, and **pollution levels**. This information is crucial for developing strategies to combat environmental challenges and helping scientists and policymakers make informed decisions about protecting our planet.
- **Education and learning** are also transforming, thanks to AI. Adaptive learning platforms tailor educational content to fit **individual student needs**, adjusting lessons based on their

progress and **learning style**. This personalization creates a more inclusive and effective learning environment and ensures that each student receives the support they need to succeed.
- AI is also making strides in **language acquisition**. Interactive tools use AI to enhance vocabulary and grammar skills, offering real-time feedback and personalized exercises. These tools adapt to the learner's pace, making the process more engaging and efficient.

Whether you are learning a new language or mastering a complex subject, AI-driven platforms offer a customized experience that caters to your unique learning style.

AI's potential to drive **social impact** cannot be overlooked. Innovative applications address societal challenges, offering solutions to issues such as **poverty**, **hunger**, and **access to education**. Imagine AI-driven systems that optimize resource distribution in underserved communities, ensuring that **food**, **healthcare**, and **educational materials** reach those who need them most.

AI can also support initiatives in **disaster response**, analyzing data to predict and mitigate the effects of natural calamities. By processing information quickly and accurately, AI models help **coordinate relief efforts,** ensuring that aid is delivered efficiently and effectively. These applications demonstrate AI's ability to enhance individual experiences and contribute to the greater good, creating a **more equitable and sustainable world**.

Multiple-Choice Questions

1. How are AI tools transforming the creation of art?

   a. By replacing human artists
   b. By creating art that can only be appreciated by machines
   c. By limiting artists to traditional techniques
   d. By assisting artists in generating unique pieces through analysis of styles

Answer: d. By assisting artists in generating unique pieces through analysis of styles

2. What role does AI play in music composition?

   a. It eliminates the need for human musicians.
   b. It generates melodies and rhythms, inspiring musicians.
   c. It composes entire albums without any human input.
   d. It only analyzes existing music without creating original compositions.

Answer: b. It generates melodies and rhythms, inspiring musicians.

3. Which of the following is a benefit of using AI in drug discovery?

   a. It guarantees the success of every new medication.
   b. It allows for faster production of medicines without research.
   c. It eliminates the need for human researchers in laboratories.
   d. It predicts interactions between compounds and biological targets to focus research.

Answer: d. It predicts interactions between compounds and biological targets to focus research.

4. How does AI contribute to environmental monitoring?

   a. By producing physical products for environmental cleanup
   b. By replacing environmental scientists with algorithms
   c. By analyzing satellite data to track climate change impacts
   d. By approving environmental policies instantly

Answer: c. By analyzing satellite data to track climate change impacts

5. What is the purpose of adaptive learning platforms powered by AI?

   a. To personalize educational experiences based on individual student needs
   b. To standardize educational content for all students
   c. To increase the number of students in traditional classrooms

d. To make learning more difficult for students

Answer: a. To personalize educational experiences based on individual student needs

6. In the context of language learning, how does AI enhance the experience?

a. By providing fixed exercises that do not adapt
b. By using only prerecorded lectures for instruction
c. By teaching multiple languages simultaneously with no focus
d. By offering real-time feedback and personalized exercises

Answer: d. By offering real-time feedback and personalized exercises

7. What potential social issues can AI help address?

a. Artistic expression
b. Resource distribution in underserved communities
c. The entertainment industry
d. Physical fitness

Answer: b. Resource distribution in underserved communities

8. How can AI assist in disaster response efforts?

a. By predicting and analyzing data to coordinate relief efforts
b. By eliminating the need for disaster readiness plans
c. By replacing first responders with robots
d. By generating social media posts regarding disasters

Answer: a. By predicting and analyzing data to coordinate relief efforts

9. What aspect of AI-driven tools makes them appealing for individual enhancement?

a. They produce generalized outputs that apply to everyone.
b. They require extensive training before use.
c. They offer customized experiences catering to user preferences.

d. They provide a purely academic approach without practical application.

Answer: c. They offer customized experiences catering to user preferences.

10. Why is it important to consider AI's social impact?

a. Because it only benefits commercial enterprises
b. Because it addresses societal challenges and promotes equity
c. Because it replaces traditional societal structures
d. Because it focuses solely on individual experiences

Answer: b. Because it addresses societal challenges and promotes equity

Embrace the process, celebrate your progress, and look forward to the exciting developments that lie ahead. With this foundation, you're well-prepared to explore the next chapter, where we'll discuss how AI can be **saved**, **shared**, and **collaborated on**, opening even more doors to innovation and discovery.

8

SHARING AND COLLABORATING ON AI PROJECTS

Let's say you've just completed your first AI model, and you are very happy with it. But what happens next? The real magic of AI begins when you share your creation, allowing others to play with it, refine it, and even build upon it. By sharing and collaborating on AI projects, you can show off your work as well as join a vast community of innovators and creators eager to learn and contribute. By preserving and exporting your AI models, you ensure that your hard work can be revisited, improved, and applied in new contexts, much like a musician sharing sheet music for others to interpret and perform.

8.1 SAVING AND EXPORTING YOUR AI MODELS

Preserving your AI models is crucial for several reasons.

First, it **increases efficiency** by avoiding the need for retraining, which can be both time-consuming and resource-intensive.

Once you've trained a model, you want to keep its unique behavior intact, preserving the nuances that make it valuable. This consistency is vital when deploying models across different systems, ensuring they perform predictably without the variations that retraining might introduce.

Additionally, saved models can be shared and reused within teams or with the wider AI community, contributing to collaboration and innovation.

Version control is important aspect of model preservation. It allows you to track the evolution of your model over time, recording each iteration and the changes made. This is especially important when collaborating with others, as it clearly records how a model has developed. Version control keeps a detailed history of your model's progression, making sure no drafts are lost.

Additionally, **data backup** ensures that your models are safeguarded against technical failures, protecting your work from unexpected losses.

When saving AI models, choosing the right file format is essential:

- **Python's Pickle** module is a popular choice for **serialization**, converting models into byte streams that can easily be stored and retrieved. This format is ideal for quick saves and loads during the development process.
- For **interoperability across platforms**, the **ONNX format** shines. It allows models to be exchanged between different AI frameworks, ensuring compatibility and flexibility.
- **HDF5 files** are excellent for storing large models with complex structures, providing efficient data management and access.

Each of these formats has its strengths, and selecting the right one depends on your specific needs and future applications.

Exporting Your Models

Exporting models for deployment involves preparing them for use in external applications. This might include converting models for web applications, where they're deployed using **APIs** to provide online predictions.

Mobile deployment is another exciting avenue, allowing you to bring AI capabilities to apps that run on **smartphones and tablets**. This requires exporting models in a way that's optimized for mobile environments, balancing performance and resource constraints.

By understanding these processes, you can effectively share your models with users, providing AI-powered features that enhance their experience. To manage your saved models, consider using tools designed for versioning and life-cycle management:

- **Data version control (DVC)** is a powerful system that tracks changes in models and datasets, ensuring that your work is organized and reproducible. It integrates seamlessly with Git, providing a familiar interface for those who use version control in software development.
- **MLflow** is another comprehensive tool that manages the entire life cycle of machine learning models, from experimentation to deployment. It offers features for tracking experiments, packaging code, and sharing results, making it an invaluable asset for collaborative AI projects.

Multiple-Choice Questions

1. Why is preserving AI models considered crucial after they are trained?

a. To ensure the model can be forgotten and retrained without concern
b. To avoid the time-consuming and resource-intensive retraining process
c. To increase the model's complexity after deployment
d. To enable models to perform differently across various systems

Answer: b. To avoid the time-consuming and resource-intensive retraining process

2. What is the primary advantage of version control in model preservation?

a. It increases the model's training time.
b. It allows tracking the evolution and changes made to the model over time.
c. It simplifies the model without saving past iterations.
d. It encourages experimentation without keeping records.

Answer: b. It allows tracking the evolution and changes made to the model over time.

3. Which file format is commonly used for quickly saving and loading AI models during development?

a. HDF5
b. ONNX
c. JSON
d. Pickle

Answer: d. Pickle

4. What is the main benefit of using the ONNX format for AI models?

a. It ensures interoperability across different AI frameworks.
b. It is only compatible with Python.
c. It simplifies the complexity of model-training processes.
d. It is used solely for visualization purposes.

Answer: a. It ensures interoperability across different AI frameworks.

5. What is typically necessary when deploying AI models for web applications?

a. Deleting the model after training is completed
b. Restricting access to only internal teams
c. Training new models from scratch for each deployment
d. Converting models for deployment using APIs

Answer: d. Converting models for deployment using APIs

6. What kind of environments can benefit from the mobile deployment of AI models?

a. Only desktop computers
b. Laptops
c. Smartphones and tablets
d. High-performance servers

Answer: c. Smartphones and tablets

7. Which tool is specifically designed for versioning models and datasets?

a. MLflow
b. TensorFlow
c. DVC
d. Scikit-learn

Answer: c. DVC

8. What capabilities does MLflow provide in managing machine learning models?

a. It solely tracks the training process without deployment features.
b. It manages the entire model life cycle, from experimentation to deployment.
c. It automatically generates data for training without user input.
d. It focuses only on visualization and reporting.

Answer: b. It manages the entire model life cycle, from experimentation to deployment.

9. What is a common challenge people face when exporting models for mobile applications?

a. Balancing performance with resource constraints
b. Ensuring the model runs on desktop devices
c. Making models inaccessible to users
d. Training models without user data

Answer: a. Balancing performance with resource constraints

10. What metaphor is used to describe sharing AI models in the context of collaboration?

a. Sharing sheet music for others to interpret and perform
b. Sharing a recipe among friends
c. Lending a book to a colleague

d. Posting a photo online

Answer: a. Sharing sheet music for others to interpret and perform

Interactive Exercise: Practice Saving and Exporting Models

To get hands-on experience:

Save a simple model using Python's Pickle module.
Then, explore exporting the same model using the ONNX format for interoperability.
Consider how these formats might be used in different scenarios, such as deploying a web application or sharing with a colleague.

This exercise will reinforce your understanding of model preservation and prepare you for future collaborations.

8.2 SHARING YOUR AI WORK WITH THE COMMUNITY

Sharing your AI projects with the community is a powerful step that opens doors to feedback, collaboration, and even new opportunities. When you put your work out there, you invite constructive feedback from peers who can offer fresh perspectives and insights. This feedback loop can be invaluable, helping you refine your models, spot areas for improvement, and build on your strengths.

Sharing your work also builds **your portfolio**—a tangible showcase of your expertise and accomplishments. A well-documented project can speak volumes about your skills, making you stand out in the AI community or when seeking career opportunities.

A **diverse portfolio** demonstrates your ability to tackle different challenges and displays the range of your capabilities. It's a dynamic, eye-catching way to show your growth in the field and maybe even serve as a springboard for your career.

Several platforms facilitate sharing AI work, each offering unique benefits:

- **GitHub** is a popular choice for publishing **code repositories**, providing an open space for collaboration and community access. Here, you can share your projects, receive feedback, and even contribute to others' work. It's a vibrant ecosystem that thrives on collaboration and shared learning.
- **Kaggle**, known for its data science competitions, offers another avenue for sharing. By participating in competitions and sharing kernels, you expose your work to a global audience of data enthusiasts, gaining insights and inspiration from diverse approaches and solutions.
- Medium and **personal blogs** are also effective for reaching broader audiences. Writing about your projects shows off your work and shares your learning experience, helping others on their path and enriching the community with your insights.
- Contributing to **open-source projects** is another rewarding way to engage with the AI community. It starts with choosing the right **open-source license**, which dictates how others can use and modify your work. Selecting a license that aligns with your goals ensures that your contributions are used ethically and in line with your intentions.
- **Documenting code** thoroughly is equally important. Clear documentation makes it possible for others to understand and build upon your work, encouraging a collaborative spirit and expanding the impact of your contributions.

Examples of successful project sharing abound, illustrating the amazing power of collaboration. Consider a project where AI was used to predict environmental changes, gaining momentum due to its open-source nature. By sharing the model, the developers invited enhancements and adaptations that led to widespread adoption in environmental research.

Another example is a collaborative project where developers worldwide contributed to improving a natural language processing tool. Their collective efforts resulted in a tool that set industry standards and influenced research practices globally. These stories underscore the potential of open collaboration, highlighting how shared projects can lead to innovation and recognition and can make an impact.

Multiple-Choice Questions

1. What is a key benefit of sharing your AI projects with the community?

a. It guarantees immediate job offers.
b. It eliminates the need for documentation.
c. It allows for constructive feedback and new perspectives.
d. It ensures the model will be used commercially.

Answer: c. It allows for constructive feedback and new perspectives.

2. How does sharing your AI work contribute to your professional portfolio?

a. It provides a record of your expertise and accomplishments.
b. It decreases your visibility in the community.
c. It limits your opportunities for collaboration.
d. It shows only your most recent projects.

Answer: a. It provides a record of your expertise and accomplishments.

3. Which platform is primarily known for publishing code repositories and encouraging collaboration?

a. Medium
b. Kaggle
c. GitHub
d. Google Docs

Answer: c. GitHub

4. What is Kaggle's primary focus in the context of sharing AI work?

a. Providing a space for personal blogs
b. Selling AI software licenses
c. Offering cloud storage for AI models
d. Hosting data science competitions and sharing kernels

Answer: d. Hosting data science competitions and sharing kernels

5. Why is thorough documentation of code important when sharing AI projects?

a. It makes the code more difficult for others to understand.
b. It ensures that others can comprehend, use, and build upon your work.
c. It prevents others from accessing your code.
d. It allows for personal notes without public visibility.

Answer: b. It ensures that others can comprehend, use, and build upon your work.

6. What role does an open-source license play in sharing AI projects?

a. It dictates how others can legally use or modify your work.
b. It ensures that your project remains completely private.
c. It prevents any modifications to your code.
d. It limits the audience to corporate users only.

Answer: a. It dictates how others can legally use or modify your work.

7. What can be a positive outcome of collaborating on AI projects?

a. Reduced need for new ideas
b. Innovation and recognition within the AI community
c. Decreased motivation to work on future projects
d. Isolation from other developers

Answer: b. Innovation and recognition within the AI community.

8. What is one example of a successful project sharing described in the content?

a. A model for predicting stock prices
b. An AI art generation tool with no community input
c. A project solely focused on individual contributions
d. A collaborative natural language processing tool developed worldwide

Answer: d. A collaborative natural language processing tool developed worldwide

9. How can writing about your AI projects in Medium or personal blogs benefit others?

a. It displays only your successes without challenges.
b. It limits the audience to your immediate network.
c. It enriches the community by sharing insights and learning experiences.
d. It eliminates the need for software documentation.

Answer: c. It enriches the community by sharing insights and learning experiences.

10. What can happen when AI models are shared with the community?

a. They become obsolete.
b. They may receive enhancements and adaptations that lead to wider adoption.
c. They only serve the original creator's intent.
d. They require constant recalibration by the owner.

Answer: b. They may receive enhancements and adaptations that lead to wider adoption.

8.3 COLLABORATING ON AI PROJECTS: TEAMWORK IN ACTION

Collaboration in an AI project is like asking a bunch of friends to help you solve a complex puzzle that has been vexing you. Each person who joins you will have different skills and perspectives; working together, you make magic happen.

Teamwork not only enhances creativity but also accelerates the innovation process. When individuals with diverse skills and backgrounds collaborate, they bring unique viewpoints that can spark new ideas and solutions. This diversity in thought is invaluable in AI, where creativity and innovation often lead to breakthroughs.

Another advantage of collaboration is the **division of labor**. Complex AI projects can be overwhelming if tackled alone, but when a team divides the tasks, the workload becomes manageable. Each team member can focus on their personal strengths, whether it's data preprocessing, model training, or deployment, allowing for a more efficient workflow.

By distributing tasks based on expertise, teams can tackle complex projects more effectively and achieve results faster.

To pave the way to smooth collaboration, several tools have become indispensable in the world of AI teamwork:

- Platforms like **Slack** and **Microsoft Teams** are crucial for communication, offering a central hub where team members can share updates, discuss ideas, and coordinate efforts. These tools help make sure that everyone stays on the same page and reduces the chances of miscommunication.
- For coding and development, **JupyterHub** provides a collaborative environment where teams can work on shared notebooks. This setup allows multiple users to interact with the same codebase, making collaborating on data analysis and model development easier.
- Meanwhile, project management tools like **Trello** and **Asana** help organize tasks and timelines, ensuring that projects stay on track and meet deadlines.

Effective collaboration also requires clear strategies and structures:

- **Regular meetings** are vital, as they provide a platform for discussing progress, addressing challenges, and aligning on goals. These check-ins can be daily, weekly, or as many as the project needs, but **the key is consistency**.
- **Role definition** is another crucial element. By **clearly assigning roles and responsibilities**, teams can avoid confusion and ensure that everyone knows their part in the project. This clarity helps prevent overlap and ensures that all aspects of the project are covered.

- Additionally, using **version control systems** like **Git** is essential for managing code changes. It allows team members to work on different parts of a project simultaneously without risking conflicts or data loss. These systems track changes, making it easy to revert to previous versions if needed.

Numerous examples of successful AI collaborations highlight the power of teamwork. Consider the development of autonomous vehicles, where teams of engineers, data scientists, and software developers work together to create sophisticated models that can navigate real-world conditions. Team members brought a variety of expertise to the table, contributing to the project's success.

Similarly, in healthcare, collaborative projects have led to the development of AI tools that assist in diagnosing diseases. These projects often involve data scientists, medical professionals, and AI researchers, each playing a role in creating solutions that can save lives. Such examples underscore the potential of collaboration to drive innovation and achieve outstanding results in AI.

Multiple-Choice Questions

1. What is a significant benefit of collaboration in AI projects?

   a. It guarantees success without effort.
   b. It reduces the need for communication tools.
   c. It makes teamwork unnecessary.
   d. It enhances creativity and speeds up the innovation process.

Answer: d. It enhances creativity and speeds up the innovation process.

2. How does dividing tasks among team members benefit AI projects?

   a. It complicates the workflow.
   b. It makes successful completion more difficult.
   c. It allows individuals to focus on their strengths and manage workloads effectively.

It ensures that every team member must perform every task.

Answer: c. It allows individuals to focus on their strengths and manage workloads effectively.

3. Which platform is mentioned as crucial for communication in AI teamwork?

a. Photoshop
b. Slack
c. GitHub
d. Jupyter Notebook

Answer: b. Slack

4. What functionality does JupyterHub provide for AI project teams?

a. It allows collaborative work on shared notebooks.
b. It offers cloud storage for large datasets.
c. It manages project finances.
d. It creates visualizations without coding.

Answer: a. It allows collaborative work on shared notebooks.

5. Why are regular meetings important in collaborative AI projects?

a. They waste time and resources.
b. They ensure that only one team member communicates updates.
c. They eliminate the need for version control.
d. They provide a platform to discuss progress and align on goals.

Answer: d. They provide a platform to discuss progress and align on goals.

6. What is the purpose of role definition in a team setting?

a. To allow everyone to take on multiple roles at once
b. To keep tasks ambiguous
c. To avoid confusion and ensure everyone knows their part in the project
d. To create competition among team members

Answer: c. To avoid confusion and ensure everyone knows their part in the project

7. How does version control, such as Git, support collaborative work?

a. It tracks changes and allows simultaneous work on different parts of a project.
b. It prevents any code changes from being made.
c. It guarantees that all team members will always have the same version.
d. It makes collaboration more challenging by complicating changes.

Answer: a. It tracks changes and allows simultaneous work on different parts of a project.

8. Which of the following is an example of successful AI collaboration?

a. Developing algorithms in isolation
b. Creating autonomous vehicles through teamwork among different specialists
c. Writing a research paper alone
d. Developing a mobile app without input from others

Answer: b. Creating autonomous vehicles through teamwork among different specialists

9. What roles do medical professionals play in collaborative AI projects in healthcare?

a. They only provide funding for the projects.
b. They completely replace data scientists.
c. They contribute their expertise to assist in diagnosing diseases.
d. They only handle administrative tasks.

Answer: c. They contribute their expertise to assist in diagnosing diseases.

10. Why is having diverse skills in a collaborative team advantageous?

   a. It leads to faster decision-making without discussion.
   b. It provides multiple perspectives that can spark new ideas and solutions.
   c. It creates conflict among team members.
   d. It makes consensus more difficult to reach.

Answer: b. It provides multiple perspectives that can spark new ideas and solutions.

With a solid foundation in collaboration and sharing, you're now ready to look at the next chapter, where we'll explore the ethical considerations of AI, ensuring that your work is not only innovative but also responsible and impactful.

9

ETHICAL AI AND RESPONSIBLE USE

Imagine stepping into a busy marketplace filled with vendors from all walks of life. Each stall represents a different aspect of AI, offering a diverse array of technologies and applications. However, beneath the surface, ethical considerations weave through every interaction, shaping the marketplace and influencing each transaction.

Ethics play a crucial role in ensuring that AI technologies serve the greater good rather than causing harm or perpetuating inequality.

9.1 UNDERSTANDING AI ETHICS: WHAT YOU NEED TO KNOW

At the heart of ethical AI are core principles:

- **Fairness** is one such principle, ensuring that AI systems do not discriminate against any group. People are creating **inclusive technologies** that respect diversity and strive for **equality**. Imagine an AI hiring tool that evaluates candidates solely based on merit, free from biases related to gender, race, age, or background. This level of fairness paves the way for a more equitable future where opportunities are accessible to everyone.

- **Transparency**, another key principle, ensures that AI processes and decisions remain understandable to users; they deserve to know how decisions are made and what data influences them.
- Lastly, **accountability** holds developers and companies responsible for AI outcomes, emphasizing the importance of ethical responsibility in technological advancement.

Despite these guiding principles, ethical dilemmas often arise:

- **Autonomous decision-making** presents a challenge, as these systems make choices without human intervention. Consider self-driving cars making split-second decisions during unforeseen road conditions. The **moral implications** of these decisions—who or what is prioritized in a potential accident—pose significant ethical questions.
- **Data ownership** is another dilemma, raising questions about who actually owns the data used by AI systems. Is it the individuals who generate the data, the companies that collect it, or the systems that analyze it? These questions highlight the complexities of data rights and privacy in an increasingly digital world.

Ethical frameworks offer a structured approach to working through these challenges. The **IEEE Global Initiative on Ethics of Autonomous and Intelligent Systems** provides comprehensive guidelines that prioritize human rights, well-being, accountability, transparency, and misuse awareness. These guidelines emphasize enhancing human well-being while safeguarding against harm and discrimination.

The **Asilomar AI Principles**, developed by the **Future of Life Institute**, also play a significant role in guiding ethical AI development. They stress the importance of aligning AI goals with human values and ensuring that AI benefits are shared widely. These principles advocate for safe, secure AI systems that respect human dignity, rights, and cultural diversity.

Real-world examples illustrate the ethical challenges that AI practitioners face. For autonomous vehicles, ethical considerations abound in decision-making. Engineers must program these vehicles to handle complex scenarios, balancing the safety of passengers, pedestrians, and other drivers. Each decision carries ethical weight, as the chosen course of action can have life-altering consequences.

Similarly, AI in hiring processes raises concerns about bias and fairness. **Automated recruitment tools** can inadvertently favor certain demographics if trained on biased data. This demonstrates the importance of ensuring that AI systems are built on diverse, representative datasets to promote fairness and inclusivity.

Multiple-Choice Questions

1. What is a core principle of ethical AI aimed at ensuring inclusiveness and equality?

a. Fairness
b. Transparency
c. Accountability
d. Efficiency

Answer: a. Fairness

2. Why is transparency important in AI systems?

a. It allows users to understand how decisions are made and what data influenced them.
b. It makes the system faster.
c. It guarantees that AI systems are always accurate.
d. It improves the technical performance of algorithms.

Answer: a. It allows users to understand how decisions are made and what data influenced them.

3. What ethical challenge arises from autonomous decision-making in AI systems, such as self-driving cars?

a. Difficulty in data collection
b. Enhanced user experience
c. Improved efficiency in transportation
d. Moral implications of prioritizing lives in accidents

Answer: d. Moral implications of prioritizing lives in accidents

4. What dilemma does data ownership in AI raise?

a. The speed of data processing
b. Questions about who owns the data generated by individuals
c. The cost of data storage
d. The importance of data visualization

Answer: b. Questions about who owns the data generated by individuals

5. Which ethical framework emphasizes human rights and accountability in autonomous systems?

a. The Asilomar AI Principles
b. The Belmont Report
c. The IEEE Global Initiative on Ethics of Autonomous and Intelligent Systems
d. The GDPR

Answer: c. The IEEE Global Initiative on Ethics of Autonomous and Intelligent Systems

6. What is one of the key focuses of the Asilomar AI Principles?

a. Maximizing profit from AI technologies
b. Simplifying AI development processes
c. Promoting AI exclusivity for advanced users only
d. Ensuring AI safety and alignment with human values

Answer: d. Ensuring AI safety and alignment with human values

7. Why is bias a concern in AI hiring processes?

a. It can favor certain demographics if trained on biased data.
b. It ensures only the best candidates are selected.
c. It reduces the time taken to recruit.
d. It directly increases diversity in the workplace.

Answer: a. It can favor certain demographics if trained on biased data.

8. In the context of autonomous vehicles, what ethical considerations must engineers consider?

a. The speed of response
b. Cost efficiency of the vehicle design
c. Balancing the safety of various road users in decision-making
d. Enhancing the luxury features of the vehicles

Answer: c. Balancing the safety of various road users in decision-making

9. What role does accountability hold in ethical AI development?

a. Reducing the costs of AI projects
b. Ensuring developers and companies are responsible for AI outcomes
c. Providing unlimited access to user data
d. Promoting the development of AI without oversight

Answer: b. Ensuring developers and companies are responsible for AI outcomes

10. How can the issue of fairness in AI be addressed?

a. By disregarding minority perspectives
b. By training AI systems on diverse, representative datasets
c. By making AI decisions completely automated
d. By promoting exclusivity in hiring processes

Answer: b. By training AI systems on diverse, representative datasets

> ***Reflection Section***
>
> **Consider how you would address ethical dilemmas in AI.**
>
> **What steps would you take to ensure fairness and transparency?**
>
> **Reflect on the role of ethical frameworks in guiding your decisions and how you might apply these principles in real-world scenarios.**

9.2 NAVIGATING PRIVACY AND BIAS IN AI

Protecting privacy in AI applications isn't just about keeping secrets; it's about ensuring that personal information remains confidential and is used responsibly:

- **Data anonymization** plays a crucial role in this process. It involves altering data so that individuals cannot be identified, even if the data is exposed.
- **Aggregation** and **pseudonymization** help ensure that user data remains private while allowing meaningful analysis. **Anonymization** acts as a shield, safeguarding personal details from prying eyes.
- **Consent mechanisms** are equally vital. They ensure that users control how their data is collected and used. Clear consent forms and options for users to access, modify, or delete their data are necessary—forms that outline what data is being collected, why it's needed, and how it will be used.
- To tackle privacy issues, **differential privacy** offers a promising solution. It involves adding "**noise**" to data, ensuring that individual information cannot be discerned from the dataset. Picture it as adding a layer of fuzziness that protects personal details while allowing for accurate analysis.
- **Secure multiparty computation** is another technique enabling computations on encrypted data while protecting the underlying information. This allows multiple parties to collaborate and

perform calculations without compromising privacy, like working in a secure vault where information remains locked but accessible for necessary tasks.

This transparency builds trust and allows users to make informed decisions about their personal information. Without proper consent mechanisms, companies risk violating privacy rights and **eroding user trust**, which can have far-reaching consequences for both individuals and organizations.

- **Bias in AI** is another pressing challenge. It can creep into systems through biased training data, skewing outcomes and perpetuating inequalities. For example, if an AI model is trained on data that reflects historical biases, it may replicate or even amplify those biases, which can lead to discrimination or unfair treatment, particularly among marginalized groups.
- **Algorithmic bias**, where the design of certain algorithms inherently favors one group over another, is another concern. This can happen inadvertently when algorithms are not carefully designed or tested. Addressing these biases is crucial for creating fair and equitable AI systems that serve all users.
- Reducing bias in AI requires proactive measures. **Conducting regular bias audits** can help identify and address biases within AI models. These audits involve assessing outcomes for different groups and making adjustments to ensure fairness.
- **Diverse data sourcing** is another critical strategy. AI systems need to learn from a more comprehensive and inclusive set of information. Datasets need to represent a wide range of demographics and scenarios. This diversity enriches the AI's understanding and enables it to make fairer and more accurate decisions.

Combining these techniques creates a well-rounded approach to managing privacy and bias in AI. AI developers can create powerful and ethical systems by prioritizing user privacy and actively working to reduce bias. These efforts enhance the quality and fairness of AI applications and build trust, ensuring that users feel secure and valued.

Multiple-Choice Questions

1. What is a critical concern regarding user privacy in AI applications?

   a. Ensuring data is collected as quickly as possible
   b. Keeping personal information confidential and used responsibly
   c. Making sure all data is public for transparency
   d. Storing personal information indefinitely

Answer: b. Keeping personal information confidential and used responsibly

2. What role does data anonymization play in protecting privacy?

   a. It makes data easier to identify individuals.
   b. It ensures data remains completely accessible to all users.
   c. It alters data so individuals cannot be identified.
   d. It deletes all data before analysis.

Answer: c. It alters data so individuals cannot be identified.

3. Why are consent mechanisms important in AI applications?

   a. They allow companies to share data without restrictions.
   b. They offer complete anonymity without user knowledge.
   c. They prevent users from accessing their own data.
   d. They enable users to control how their data is collected and used.

Answer: d. They enable users to control how their data is collected and used.

4. How can bias enter AI systems?

   a. From biased training data that reflects historical biases
   b. Through universally accepted training data
   c. By using standardized algorithms without variation
   d. From the random selection of data inputs

Answer: a. From biased training data that reflects historical biases

5. What is differential privacy?

a. A method to completely eliminate all user data
b. A technique that adds noise to data to protect individual information
c. A way to make data publicly available
d. A technique to merge data from different users

Answer: b. A technique that adds noise to data to protect individual information.

6. What does secure multiparty computation allow?

a. Sharing sensitive data openly among parties
b. Storing information in a public cloud
c. Performing computations on encrypted data without revealing it
d. Combining data from multiple sources without security

Answer: c. Performing computations on encrypted data without revealing it

7. How can regular bias audits help in AI development?

a. They ensure all data is ignored.
b. They eliminate the need for diverse datasets.
c. They reduce the complexity of algorithms.
d. They identify biases within AI models and allow for adjustments.

Answer: d. They identify biases within AI models and allow for adjustments.

8. What is the benefit of diverse data sourcing in AI?

a. It ensures that datasets represent a wide range of demographics and scenarios.
b. It simplifies the training process for models.
c. It guarantees that no biases can occur.
d. It increases the speed of data processing.

Answer: a. It ensures that datasets represent a wide range of demographics and scenarios.

9. What is the ultimate goal of combining privacy protection and bias reduction techniques in AI?

a. To eliminate all AI applications
b. To create powerful and ethical AI systems that users trust
c. To limit the use of AI to only large corporations
d. To encourage users to share their personal data freely

Answer: b. To create powerful and ethical AI systems that users trust

10. What metaphor is used to describe the importance of diverse data in AI?

a. A blank canvas waiting to be painted
b. A single path through the forest
c. A diverse library of books offering unique stories and perspectives
d. A crowded market with identical stalls

Answer: c. A diverse library of books offering unique stories and perspectives

9.3 IMPLEMENTING RESPONSIBLE AI PRACTICES

When discussing creating and using AI responsibly, we're referring to a development approach that prioritizes ethics and societal well-being at every stage. Think of it as building a house: You need a solid foundation of ethical guidelines to ensure the structure is both safe and functional.

These guidelines act like blueprints, helping you navigate complex decisions and ensuring your AI projects align with broader societal values. They include principles like fairness, respect, and accountability, an integral part of every model and application.

Engaging diverse groups in the development process is crucial, akin to gathering perspectives from various experts before constructing a building. This inclusion ensures that AI systems are designed with a holistic view, considering the needs and impacts on different communities. By involving stakeholders from various backgrounds, you can develope AI solutions that are not only innovative but also equitable and inclusive.

The Importance of Regulation

Regulation plays a pivotal role in maintaining ethical standards in AI, much like building codes ensure the safety of physical structures. These frameworks provide the legal backbone that guides the ethical development and deployment of AI.

The **General Data Protection Regulation (GDPR)** is a prime example, setting stringent guidelines for data protection and privacy. Organizations need to handle personal data responsibly, emphasizing transparency and user consent. This regulation impacts AI systems by enforcing accountability and ensuring that data-driven decisions respect individual rights.

Emerging AI-specific legislation further reinforces these principles, focusing on ethics and accountability in technological advancement. These laws aim to balance innovation with protection, safeguard against misuse, and ensure that AI serves the public good.

Leading companies in the AI industry have adopted best practices to ensure responsible use of their technologies. For instance, many have established **AI ethics boards**, internal committees that oversee ethical considerations in AI projects. These boards act as guardians of ethical integrity, reviewing projects and providing guidance to ensure alignment with ethical standards.

Regular ethical reviews are also a common practice, integrated into the development lifecycle to continuously assess and improve ethical compliance. These reviews serve as checkpoints, ensuring that each development phase adheres to ethical guidelines and addresses potential issues before they escalate.

By adopting these practices into their operations, companies can develop a culture of ethical awareness and accountability.

For those seeking to implement ethical AI, numerous tools and resources are available to guide the process. These resources offer practical support, from frameworks and guidelines to training programs and workshops. They serve as invaluable aids, helping developers and organizations navigate the complexities of ethical AI implementation:

- Tools like **AI impact assessments** provide structured approaches to evaluating the potential effects of AI systems, ensuring they align with ethical standards and societal values.
- Additionally, resources such as **ethical AI toolkits** offer comprehensive guidance on integrating ethics into the design and deployment of AI technologies. These toolkits include checklists, case studies, and best practices.

To truly embrace responsible AI, viewing it as a **continuous commitment** rather than a one-time task is important. This perspective encourages ongoing reflection and adaptation, ensuring that AI systems remain current with evolving ethical standards and societal needs.

By committing to ethical practices, you build a future where AI enhances lives, respects rights, and empowers communities. This commitment builds trust and collaboration which paves the way for solutions that address real-world challenges while upholding ethical integrity.

Multiple-Choice Questions

1. What does responsible AI development prioritize?

a. Maximizing profits
b. Innovation without oversight
c. Ethics and societal well-being
d. Competition among companies

Answer: c. Ethics and societal well-being

2. In the context of AI development, why is engaging diverse groups important?

a. It slows down the development process.

b. It ensures that various perspectives are considered, making AI solutions more equitable.
c. It complicates decision-making without adding value.
d. It allows for a focus on a single demographic.

Answer: b. It ensures various perspectives are considered, making AI solutions more equitable.

3. Which regulation sets stringent guidelines for data protection and privacy in AI?

a. Freedom of Information Act
b. Telecommunications Act
c. GDPR
d. Digital Millennium Copyright Act

Answer: c. GDPR

4. What role do AI ethics boards play in companies?

a. They oversee ethical considerations in AI projects.
b. They focus solely on profit margins.
c. They manage marketing strategies.
d. They ensure technical development only.

Answer: a. They oversee ethical considerations in AI projects.

5. What is an AI impact assessment?

a. A financial analysis of AI systems
b. A report on AI training data quality
c. A method to increase the efficiency of AI models
d. A tool for evaluating the potential effects of AI systems on ethical standards

Answer: d. A tool for evaluating the potential effects of AI systems on ethical standards

6. Which of the following describes responsible AI as a commitment?

a. A continuous process of reflection and adaptation
b. A one-time task to complete when launching a product
c. A fixed set of rules that do not change over time
d. A style of development that prioritizes speed over ethics

Answer: a. A continuous process of reflection and adaptation

7. What is the purpose of embedding regular ethical reviews in the AI development lifecycle?

a. To reduce development costs
b. To eliminate the need for diverse perspectives
c. To accelerate project completion
d. To ensure continuous assessment and improvement of ethical compliance

Answer: d. To ensure continuous assessment and improvement of ethical compliance

8. What type of resources are available to guide ethical AI implementation?

a. Financial reports
b. Frameworks, guidelines, and training programs
c. Competing product designs
d. User manuals for software

Answer: b. Frameworks, guidelines, and training programs

9. What is one example of a best practice adopted by leading AI companies?

a. Ignoring ethical considerations during development
b. Establishing AI ethics boards to oversee projects
c. Prioritizing financial gain above all else
d. Reducing collaboration with external stakeholders

Answer: b. Establishing AI ethics boards to oversee projects

10. How does responsible AI contribute to society?

a. By ignoring user rights and concerns
b. By using technology to control communities
c. By enhancing lives, respecting rights, and empowering communities
d. By focusing solely on technological advancement without ethical considerations

Answer: c. By enhancing lives, respecting rights, and empowering communities.

The journey through AI ethics requires both accumulating knowledge and developing a nuanced understanding of how these principles apply in various contexts. As you continue to explore this field, remember that the goal is not perfection but progress.

10

EXPLORING ADVANCED AI CONCEPTS

Neural networks process information similar to how the human brain works. They are built of layers of interconnected nodes, known as neurons, which are the building blocks of these networks. Each neuron receives input, processes it, and passes it along to the next layer, much like how our brain processes stimuli. This interconnected structure allows neural networks to recognize and learn from patterns, and make decisions in a powerful and adaptable way.

10.1 DIVING INTO NEURAL NETWORKS AND DEEP LEARNING

In a neural network, neurons are organized into 3 layers: the input layer, one or more hidden layers, and the output layer.

- The **input layer** receives the initial data, much like when your senses are taking in the world.
- **Hidden layers**, which can number from one to many, process this information through complex calculations, gradually abstracting and refining it.
- Finally, the **output layer** delivers the result, such as a prediction or classification.

Activation functions are crucial in determining whether or not a neuron gets activated based on the input. Common activation functions include the **sigmoid** and **ReLU** (short for Rectified Linear Unit), each with unique properties that influence how the network learns and processes data.

Deep learning extends the capabilities of neural networks by adding depth, known as **deep neural networks**. With these multiple hidden layers, the system can learn increasingly abstract features from the data.

Imagine how a child learns to recognize animals: Initially, they learn basic shapes, then colors, and finally, specific features like stripes or spots. Similarly, deep learning models layer these features, creating a hierarchy that allows the network to understand complex patterns.

Backpropagation is the process through which these networks learn from mistakes. It involves adjusting the weights of connections between neurons based on the error in predictions, a bit like fine-tuning a recipe until it's just right.

Different architectures of neural networks cater to various applications:

- **Convolutional neural networks (CNNs)** are designed to process structured grid data, such as images. They excel in recognizing visual patterns, making them invaluable for image recognition and classification tasks. For example, CNNs power facial recognition software and help autonomous vehicles identify obstacles.
- **Recurrent neural networks (RNNs)**, on the other hand, are adept at handling sequential data. They are used in applications ranging from language translation to time series analysis, where the order of information is crucial.
- **Generative adversarial networks (GANs)** take creativity to a new level, pitting two networks against each other to produce new data samples like realistic images or music, pushing the boundaries of AI-generated content.

In practical terms, deep learning has revolutionized industries by enabling breakthroughs that were once considered science fiction. In the field of autonomous driving, deep learning models are used for **object detection** and **path planning**, allowing cars to navigate safely through complex environments. These systems process vast amounts of sensor data to make real-time decisions, enhancing both safety and efficiency on the road.

In healthcare, deep learning assists in **medical imaging**, where CNNs analyze scans to detect diseases with remarkable accuracy. By identifying patterns that might be invisible to the human eye, these models support early diagnosis and treatment, ultimately improving patient outcomes.

Multiple-Choice Questions

1. What do neural networks mimic in their functionality?

   a. The way the human brain processes information
   b. The way computers store information
   c. The processes of mechanical machines
   d. The features of traditional programming

Answer: a. The way the human brain processes information

2. What is the role of neurons in a neural network?

   a. To only store data permanently
   b. To receive input, process it, and pass it to the next layer
   c. To independently make final decisions without input
   d. To output data without processing

Answer: b. To receive input, process it, and pass it to the next layer

3. What is the function of the activation function in a neural network?

   a. To prevent the network from learning
   b. To define the structure of the network
   c. To create permanent outputs without input
   d. To determine whether a neuron should be activated based on input

Answer: d. To determine whether a neuron should be activated based on input

4. How does deep learning enhance the capabilities of neural networks?

a. By adding more input layers
b. By simplifying the model to a single layer
c. By including multiple hidden layers for more complex feature extraction
d. By reducing the amount of data processed

Answer: c. By including multiple hidden layers for more complex feature extraction

5. What is the purpose of backpropagation in neural networks?

a. To fine-tune the weights of connections based on errors in predictions
b. To initialize the network for training
c. To create new neurons within the network
d. To speed up the data processing time

Answer: a. To fine-tune the weights of connections based on errors in predictions

6. What type of neural network is particularly effective for image recognition tasks?

a. RNNs
b. Decision trees
c. Feedforward neural networks
d. CNNs

Answer: d. CNNs

7. Which neural network architecture is best suited for sequential data?

a. CNNs
b. GANs
c. RNNs
d. Support vector machines

Answer: c. RNNs

8. What innovative applications do GANs support?

a. Data encryption
b. Producing new data samples, such as realistic images or music
c. Enhancing image resolution
d. Compressing large datasets

Answer: b. Producing new data samples, such as realistic images or music

9. In which field does deep learning help with analyzing medical imaging?

a. Financial forecasting
b. Autonomous driving
c. Agriculture
d. Healthcare

Answer: d. Healthcare

10. What significant benefit does deep learning provide in autonomous driving applications?

a. It reduces the number of sensors required.
b. It helps with object detection and path planning.
c. It eliminates the need for human oversight entirely.
d. It slows down reaction times to improve safety.

Answer: b. It helps with object detection and path planning.

Interactive Exercise: Reflect on Neural Networks

Consider the last time you encountered a smart device, whether it was a phone app or a smart speaker.

Consider how neural networks might have been at work, processing your voice commands or suggesting your next favorite song.

Reflect on the layers of information and decision-making that created that seamless interaction.

10.2 NATURAL LANGUAGE PROCESSING: AI AND HUMAN LANGUAGE

Think about a time when you sat at a café, eavesdropping on conversations around you. Each word and inflection held meaning, allowing you to grasp the mood, intent, and context of the speakers.

Natural language processing (NLP) is the technology that enables machines to perform similar feats with human language, allowing AI to interpret, analyze, and even generate text. Through NLP, computers can understand the nuances of language, from the syntax of sentences to the emotions conveyed in a tweet. This capability makes AI a more intuitive and responsive tool in our daily lives.

Text Analysis

One of the foundational aspects of NLP is **text analysis**, which involves extracting meaningful information from text data. As an example, let's say you have a mountain of customer reviews for a product. Text analysis can sift through this data, identifying key themes, trends, and sentiments. This process is like having an assistant who can read all the reviews and summarize the general opinions, saving you an enormous amount of time and effort.

Sentiment analysis can go a step further by determining the emotional tone behind the text. Is the feedback positive, negative, or neutral? Understanding these sentiments can guide businesses in improving their products and services, ensuring they meet customer expectations.

To achieve this level of comprehension, NLP employs several key techniques:

- **Tokenization** breaks down text into manageable pieces, such as words or sentences, making it easier to analyze. It's like taking a complex puzzle and sorting the pieces before assembling them.
- **Named entity recognition** is another crucial technique, where specific entities, like names, dates, or locations, are identified and classified within the text. This is particularly useful in applications like search engines, where determining the context of a query can vastly improve search results.

- **Part-of-speech tagging** assigns grammatical categories to words, helping AI understand the function of each word in a sentence, much like how a teacher might diagram a sentence to show its structure.

Modern NLP owes much of its advancement to sophisticated models like transformers. These models, such as **Bidirectional Encoder Representations from Transformers (BERT)** and **Generative Pre-trained Transformers (GPT)**, have revolutionized language processing by understanding context in ways traditional models couldn't.

Transformers use attention mechanisms to weigh how important different words in a sentence are, allowing them to grasp complex language structures and nuances.

- BERT, for instance, processes the importance of words in relation to all other words in a sentence, providing a deep understanding of context.
- GPT, on the other hand, excels at generating text and crafting coherent and contextually relevant content.

These models have significantly enhanced AI's ability to understand and generate human language. This makes interactions with machines feel more natural and intuitive.

In practical terms, NLP is already significantly impacting various domains. Powered by NLP, **chatbots** have become ubiquitous in customer service, handling queries, providing information, and even processing transactions. They are available 24/7, offering immediate assistance and freeing people to handle other issues.

Machine translation, another application of NLP, has seen remarkable improvements in accuracy and fluency, breaking down language barriers and allowing seamless communication across cultures. Whether traveling abroad or collaborating with international colleagues, these tools make connecting and sharing ideas easier without language getting in the way.

Multiple-Choice Questions

1. What does NLP enable machines to do?

a. Perform complex mathematical calculations
b. Engage in nonverbal communication
c. Conduct physical activities like walking or running
d. Interpret, analyze, and generate human language

Answer: d. Interpret, analyze, and generate human language

2. What is the purpose of text analysis in NLP?

a. To translate text into different languages
b. To create visually appealing graphics from text
c. To generate random text based on user input
d. To extract meaningful information from text data

Answer: d. To extract meaningful information from text data

3. How does sentiment analysis enhance text analysis?

a. By determining the emotional tone behind the text
b. By summarizing all data into one number
c. By categorizing text into positive and negative sentiments only
d. By counting the number of words in the text

Answer: a. By determining the emotional tone behind the text

4. What is tokenization in the context of NLP?

a. Analyzing the frequency of words in text
b. Breaking down text into manageable pieces, such as words or sentences
c. Generating new text based on existing data
d. Combining multiple texts into a single document

Answer: b. Breaking down text into manageable pieces, such as words or sentences

5. Which technique involves identifying and classifying specific entities within text?

a. Text summarization
b. Sentiment analysis
c. Named entity recognition
d. Syntax analysis

Answer: c. Named entity recognition

6. What advantages do models like BERT and GPT offer in NLP?

a. They revolutionize language processing by understanding context deeply.
b. They perform simple text analysis.
c. They are only used for translating small phrases.
d. They require no training data to function.

Answer: a. They revolutionize language processing by understanding context deeply.

7. How do transformers enhance NLP capabilities?

a. By simplifying the language used in text
b. By limiting the context to only a few words
c. By eliminating the need for any language training
d. By using attention mechanisms to weigh the importance of words

Answer: d. By using attention mechanisms to weigh the importance of words

8. What practical application of NLP involves handling customer queries and transactions?

a. Chatbots
b. Data mining
c. Speech recognition
d. Virtual reality

Answer: a. Chatbots

9. What is a significant benefit of machine translation as an application of NLP?

a. It creates more complex texts.
b. It breaks down language barriers and allows seamless communication.
c. It only works in specific languages.
d. It eliminates the need for international collaboration.

Answer: b. It breaks down language barriers and allows seamless communication.

10. What aspect of human language does NLP aim to replicate in its functionality?

a. The physical production of speech sounds
b. The emotional aspect of communication only
c. The complexities of understanding context and meaning
d. The artistic expression in writing

Answer: c. The complexities of understanding context and meaning

10.3 THE FUTURE OF AI: TRENDS AND INNOVATIONS

AI is evolving rapidly, and its future is filled with intriguing possibilities. One such trend is **explainable AI (XAI)**. In a world where algorithms make complex decisions, it's crucial to understand their reasoning.

XAI focuses on developing models that are able to explain their choices in a way humans can understand. Imagine a doctor using an AI system to diagnose a patient. With XAI, the doctor can see the diagnosis and the reasons behind it. This transparency builds trust and allows humans to verify and refine AI decisions, enhancing outcomes in fields from healthcare to finance.

Another trend reshaping AI is **edge AI**. Traditionally, AI processes data in centralized locations, like data centers. However, edge AI brings intelligence directly to devices at the network's edge, such as smartphones or gadgets. This shift means faster processing and real-time decision-making without relying on constant internet access. Here are some trends:

- Think of a **smart thermostat** that adjusts your home's temperature based on your habits. It reacts instantly to changes by **processing data locally**, **reducing latency**, and **improving efficiency**. This trend is crucial as more devices connect to the internet, creating a seamless, responsive experience in smart homes and cities.
- AI's impact is transforming how we work and live. In workplaces, AI is reshaping job roles and skill requirements. **Automation** handles routine tasks, freeing humans for more strategic work. This shift demands new skills like data analysis and AI management, emphasizing the need for continuous learning.
- Meanwhile, **smart cities** are emerging, where AI optimizes **urban planning**. From traffic management to energy distribution, AI-driven solutions make cities more efficient and sustainable. AI can predict traffic patterns and adjust signals to ease congestion. Such innovations enhance the quality of life, making urban living more convenient and environmentally friendly.
- In **space exploration**, AI plays a pivotal role. **Autonomous systems** guide spacecraft on missions, analyzing real-time data to make decisions without human intervention. This capability is vital for exploring distant planets where communication delays are significant. AI can process images from rovers on Mars, identifying potential sites for further investigation.
- Similarly, **quantum computing** is revolutionizing AI, solving problems too complex for classical computers. By harnessing quantum bits, or **qubits**, AI models can perform calculations at unprecedented speeds, opening new frontiers in research and development.

As AI advances, ethical and societal implications arise that demand careful consideration. The integration of AI into everyday life raises questions about privacy and security. In smart cities, for instance, collecting vast amounts of data must be managed responsibly to protect citizens' privacy.

There's also a challenge of ensure that AI systems remain unbiased and equitable. As they play a larger role in decision-making, from hiring to law enforcement, it's crucial to eliminate bias that could perpetuate inequalities. These challenges underscore the need for robust frameworks and regulations to guide AI development, ensuring it benefits society as a whole.

Multiple-Choice Questions

1. What is the primary focus of XAI?

   a. To develop models that explain their decisions in understandable terms
   b. To increase the speed of AI algorithms
   c. To reduce the complexity of AI systems
   d. To create AI that operates without any data

Answer: a. To develop models that explain their decisions in understandable terms

2. How does edge AI differ from traditional AI processing?

   a. It processes data in centralized data centers.
   b. It processes all data in the cloud.
   c. It requires constant internet connectivity for functionality.
   d. It brings intelligence directly to devices at the network's edge.

Answer: d. It brings intelligence directly to devices at the network's edge.

3. What is one significant benefit of edge AI?

   a. It allows for faster processing and real-time decision-making.
   b. It increases latency in processing.
   c. It requires more energy than traditional AI.
   d. It limits the functionality of devices.

Answer: a. It allows for faster processing and real-time decision-making.

4. How is AI changing job roles in the workplace?

a. By eliminating the need for human workers entirely
b. By making all jobs require manual labor
c. By automating routine tasks and freeing humans for strategic work
d. By reducing the amount of data generated by companies

Answer: c. By automating routine tasks and freeing humans for strategic work

5. In the context of smart cities, how does AI improve urban living?

a. By creating more traffic jams
b. By optimizing urban planning and managing resources efficiently
c. By reducing the number of people in cities
d. By increasing carbon emissions from transportation

Answer: b. By optimizing urban planning and managing resources efficiently

6. What role does AI play in space exploration?

a. It assists in analyzing data in real time and making autonomous decisions.
b. It replaces all human roles in space missions.
c. It is used only for communication between Earth and spacecraft.
d. It provides virtual reality experiences for astronauts.

Answer: a. It assists in analyzing data in real time and making autonomous decisions.

7. What is one advantage of quantum computing in AI applications?

a. It can only solve simple problems faster.
b. It operates on classical bits to process data.
c. It allows AI models to perform calculations at unprecedented speeds.
d. It reduces the need for machine learning algorithms.

Answer: c. It allows AI models to perform calculations at unprecedented speeds.

8. What ethical concerns arise with the integration of AI into everyday life?

a. Increased job opportunities in all sectors
b. The need for robust frameworks to manage data privacy and bias
c. A decrease in the reliance on technology
d. The elimination of all ethical issues in decision-making

Answer: b. The need for robust frameworks to manage data privacy and bias

9. Why is it essential to ensure that AI systems remain unbiased and equitable?

a. To enhance the company's profits
b. To decrease the complexity of AI systems
c. To limit the use of AI in sensitive areas
d. To prevent perpetuating social inequalities in decision-making processes

Answer: d. To prevent perpetuating social inequalities in decision-making processes

10. What do the trends like XAI and edge AI suggest about the future of AI?

a. AI will become less involved in real-world applications.
b. There will be a reduction in the use of AI across industries.
c. AI technologies will solely focus on increasing complexity.
d. Future AI will prioritize transparency, efficiency, and user control.

Answer: d. Future AI will prioritize transparency, efficiency, and user control.

The future of AI is bright and filled with potential that can transform industries and improve lives. AI's possibilities are vast, from making cities smarter to exploring the cosmos. However, with great power comes great responsibility. As we embrace these advancements, we must remain vigilant, guided by ethical principles that ensure AI serves humanity's best interests.

11

ENHANCING SKILLS WITH INTERACTIVE EXERCISES

Building an AI model is much like assembling a complicated puzzle: Each step brings clarity and understanding, transforming what once seemed complex into something manageable and enjoyable. Engaging with AI hands-on is a crucial part of learning that goes beyond reading or listening. By actively building and experimenting with models, you can move from theory to practice, gaining insights that can only come from doing.

This chapter will guide you through creating a simple AI model, offering a practical approach to deepen your understanding.

11.1 HANDS-ON PRACTICE: BUILDING A SIMPLE AI MODEL

Let's start with a classic problem: **predicting house prices**. It's an excellent choice for beginners because it involves clear, measurable variables and provides immediate, tangible results. We'll use a basic **linear regression model**, a fundamental technique in AI that predicts a numerical outcome based on input variables.

Hands-on Practice: Building a Simple AI Model for Predicting Housing Prices

For this exercise, the Boston Housing dataset is an ideal choice. It's small enough to manage easily yet rich enough to provide meaningful insights. You'll find data on various factors like the number of rooms, crime rate, and proximity to employment centers, all of which influence housing prices. This dataset serves as a perfect playground for practicing data handling and model building.

Now, on to the practical implementation, Python is your best friend here because of its simplicity and powerful libraries. We'll use Scikit-learn, a popular library that makes building machine-learning models straightforward.

1. Begin by loading and exploring your dataset to understand the data types and distributions.
2. Next, **preprocess the data**—handle missing values and normalize the data to ensure consistency.
3. With your data ready, define your **linear regression model** in Python. This involves fitting the model to your data, essentially teaching it to recognize patterns between the input variables and the house prices.
4. Finally, evaluate the model's performance using metrics like mean squared error to ensure your predictions are on point.

Hands-on practice is invaluable because:

- It solidifies theoretical concepts by putting them into action, making the abstract tangible.
- You break down complex tasks into manageable parts through **coding exercises**, reinforcing your understanding with each line of code you write.
- **Debugging**, though challenging, is an integral part of this process. It teaches you to identify and fix errors, which sharpens your problem-solving abilities.

The more you practice, the more proficient you become, turning challenges into opportunities for growth. As you work through building your model, you'll undoubtedly face obstacles. Errors in code, unexpected results, or performance issues are common hurdles. But fear not—these challenges are part of the learning process:

- Develop strategies for error handling, such as reading error messages carefully and using print statements to track variables.
- Manage your resources wisely, using efficient coding practices. Ensure your computer doesn't slow down by closing unnecessary applications.
- Remember, every problem you solve boosts your confidence and prepares you for more complex projects.
- Reflection is a powerful tool in learning. Take time to review your process, noting both successes and setbacks.
- Journaling your experiences can be insightful, helping you track your progress and identify patterns in your learning.
- Sharing your projects with peers can also be incredibly beneficial. Their feedback provides new perspectives and ideas, often highlighting aspects you might have missed.
- Embrace these opportunities for collaboration, as they enrich your learning experience and expand your understanding.

Call-to-Action: Share and Reflect

Take a moment to consider your project's impact:

Share your application with a friend or colleague and ask for their feedback. How does their perspective differ from yours?
Reflect on their suggestions and consider how to incorporate their insights into improving your model.

This practice strengthens your project and enriches your understanding of AI, making your learning experience more dynamic and rewarding.

11.2 SELF-ASSESSMENT: TESTING YOUR AI KNOWLEDGE

Imagine what its like to learn to play a musical instrument. You practice scales, learn new pieces, and now and then, you play a short recital for yourself to see how far you've come. Self-assessment in AI is much like this.

It's about knowing the right notes and understanding how they fit together to create harmony and melody. Self-assessment helps you monitor your progress, identify areas that need more attention, and boost your confidence as you see improvement over time.

There are plenty of tools and resources available to support self-assessment in AI. To make the most of self-assessment, having a structured framework is helpful:

- Start with **checklists** that cover the topics you've studied. Checking off items as you go provides a sense of accomplishment and ensures you don't miss any critical concepts.
- **Flashcards** are excellent tools for self-assessment; you can use the power of spaced repetition to reinforce learning. With digital tools like **Anki**, you can create custom flashcards focusing on specific topics, allowing you to revisit them periodically. This method helps embed knowledge into your long-term memory, which makes it easier to recall when needed. Think of flashcards as your personal AI glossary, a quick reference that grows with your understanding. They're handy for memorizing terms, algorithms, or processes foundational to your AI studies. Flipping through flashcards can turn a quick coffee break into a productive review session, seamlessly integrating learning into your daily routine.
- **Reflection questions** can further enhance this process by prompting deeper understanding and critical thinking. Ask yourself why certain algorithms work better for specific problems or how a particular AI model could be improved. These questions encourage you to connect the dots and see the bigger picture, fostering a more comprehensive grasp of AI principles. Reflecting on your answers also highlights areas where you might need additional study, guiding your future learning efforts.
- **Online platforms** offer quizzes and challenges that test your skills in various areas, from basic concepts to advanced applications.
- Participating in AI programming **contests**, like those on Kaggle, provides a real-world context for your learning, allowing you to apply concepts in a competitive yet supportive environment.

These contests test your skills and expose you to diverse approaches and solutions, enriching your understanding and sparking new ideas.

Whether through quizzes, flashcards, or interactive challenges, self-assessment is vital to your AI learning toolkit, helping you grow and succeed in your studies.

11.3 PROJECT-BASED LEARNING: CREATE YOUR AI APPLICATION

Think of **project-based learning** as the playground where theory comes to life. It's where you blend different AI concepts into a practical, cohesive whole, transforming abstract ideas into something tangible.

Creating a project allows you to tackle real-world challenges head-on, bringing together the different facets of AI development from inception to deployment. This approach is invaluable because it mimics real-world scenarios where problems are rarely isolated. Instead, they require a blend of skills and knowledge to solve effectively. By selecting a project that addresses a practical issue, you engage with AI on a deeper level, applying what you've learned in meaningful ways.

For example, you might choose to develop a sentiment analysis tool for social media. This project doesn't just teach you about NLP; it also provides insights into how data from platforms like X (formerly Twitter) can be used to gauge public opinion.

Comprehensive Learning Project: Project-Based Learning

Choosing the right project is crucial. It should align with your interests and skill level, ensuring it's both engaging and attainable.

1. Select a project that addresses a practical issue.
2. Begin by outlining your objectives—what do you hope to achieve with your project?
3. Set a timeline that includes all the stages of development, from data collection to model deployment.

4. Consider the resources you'll need, such as access to data and computing power.
5. Start with data preparation, ensuring your dataset is clean and ready for analysis. This involves checking for missing values, normalizing data, and splitting data into training and testing sets.
6. Next, build your model, select the right algorithms to process your data, and make predictions.
7. Once your model is built, evaluate its performance using metrics to identify areas for improvement.
8. Deployment is the final phase, where your application goes live, ready to interact with real users. But the work doesn't stop there.

Iterative improvement is key. Use feedback from users to refine your model, making it more accurate and efficient over time.

Whether you present your project on a platform like GitHub or share it with a local AI group, the feedback you receive can be invaluable. It highlights areas you might have overlooked and introduces you to different approaches and techniques. Additionally, collaborating with peers enhances your learning and builds a sense of community. It shows that AI isn't just a solitary pursuit but a collaborative effort that benefits from diverse viewpoints.

Call-to-Action: Share and Reflect

Take a moment to consider your project's impact:

Share your application with a friend or colleague and ask for their feedback. How does their perspective differ from yours?

Reflect on their suggestions and consider how to incorporate their insights into improving your model.

This practice strengthens your project and enriches your understanding of AI, making your learning experience more dynamic and rewarding.

As you wrap up this chapter, reflect on the various tools and techniques you've explored to reinforce your learning. Embrace these resources as you continue to build your skills, knowing that each step forward brings you closer to mastering the complexities of AI.

In the next chapter, you'll read about the diverse applications of AI across different industries, seeing firsthand how the concepts you've learned come to life in the real world.

12

REVISITING AI'S IMPACT IN SEVERAL INDUSTRIES IN MORE DEPTH

AI is reshaping patient care, offering tools that enhance diagnostic accuracy and accelerate the development of personalized treatments. We are starting to get a glimpse of AI's life-changing role in healthcare, whether it's analyzing complex medical images or providing real-time patient support. AI is revolutionizing how we approach healthcare, making it more precise and effective.

12.1 AI IN HEALTHCARE: TRANSFORMING PATIENT CARE

In diagnostics, AI is an invaluable assistant to doctors by improving the accuracy and speed of diagnoses. AI systems can analyze X-rays, MRIs, and CT scans with incredible precision, often surpassing human capabilities.

- In chest X-rays, for instance, AI algorithms can detect subtle patterns that might be missed by the human eye. This can greatly improve the ods for early detection of diseases like lung cancer or even COVID-19 (Gudiger et al., 2021; National Cancer Institute, 2024). This allows for quicker diagnosis and opens the door for timely intervention, improving patient outcomes significantly.

- **Predictive analytics** marks another leap forward in healthcare, using patient data to foresee potential health risks. By analyzing vast amounts of information, AI can identify trends and predict disease susceptibility, allowing personalized preventive care plans.
- AI's role in genetic analysis is particularly groundbreaking, offering insights from **genomic data** that lead to customized treatment plans designed for an individual's needs. This personalization also extends to **drug discovery**, where AI simulations accelerate the development of new medications by predicting their effects and interactions, drastically reducing the time and cost traditionally required for drug development.
- Regarding patient management, AI is making significant strides in enhancing the healthcare experience. **Virtual health assistants** powered by AI are available 24/7 to provide patients with information and support. This constant availability ensures that patients receive the help they need when they need it, without long wait times.

Early Detection in Radiology

One notable case of the use of AI in early disease detection is in the field of radiology. AI systems designed to **analyze mammograms** have significantly reduced radiologist workload while maintaining high diagnostic accuracy. AI tools can reduce mammograms' reading time. Radiologists can then focus on more complex cases and improve efficiency. This combination of AI's analytical prowess and human expertise is proving to be a powerful force in advancing healthcare.

AI also optimizes scheduling, ensuring that appointments and resources are allocated efficiently, minimizing delays and enhancing the workflow within healthcare facilities.

To illustrate AI's transformative power in healthcare, consider **IBM Watson Health**, a system that assists clinicians in making more informed decisions by analyzing extensive medical data to provide evidence-based treatment recommendations.

Similarly, **Google's DeepMind** has been a pioneer in predicting patient deterioration in hospitals, enabling early intervention and improved patient care. These examples show how AI supports healthcare professionals and enhances the standard of care provided to patients.

As AI evolves, its integration into healthcare promises to enhance patient care and transform the entire industry, making it more responsive, efficient, and adaptable to individual needs.

12.2 AI IN FINANCE: REVOLUTIONIZING BANKING AND INVESTMENTS

Imagine walking into your bank where the teller knows your preferences and needs, and is ready to offer personalized services that fits your lifestyle. That's the magic of AI in finance, where efficiency, security, and personalization are reshaping the industry.

- AI systems are like vigilant guardians, identifying real-time suspicious transactions to **prevent fraud**. This isn't just about catching a thief in the act but about safeguarding your financial peace of mind. By analyzing patterns that may hint at fraudulent activity, AI protects your assets and ensures that genuine transactions proceed smoothly.
- Meanwhile, AI is making credit scoring more accurate and fair in **lending**. By evaluating a broader set of data points, AI can provide a more nuanced and equitable assessment of **creditworthiness**, opening doors for those who might have been overlooked by traditional methods.
- In the fast-paced world of **investment and trading**, AI is a game-changer. Picture a bustling stock exchange with AI algorithms tirelessly analyzing market data at lightning speed. This is the heart of algorithmic trading, where AI models execute high-frequency trades based on subtle market shifts that humans might miss. These algorithms process endless amounts of information and can make split-second decisions that can lead to significant gains.
- Beyond trading, AI powers portfolio management tools that offer personalized investment advice. These tools suggest

diversification strategies personalized to individual risk profiles and continuously adapt to market changes, ensuring that investments remain aligned with financial goals.

- **Customer service** in banking has also seen a revolution, thanks to AI. Gone are the days of waiting on hold for simple inquiries. AI-driven chatbots are now the frontline of customer service, handling routine questions easily and precisely. These virtual assistants provide instant responses, reduce wait times, and free up human agents for more complex issues.
- AI also dives deep into customer data to offer personalized financial products. Whether suggesting a savings plan or a loan with favorable terms, AI analyzes individual financial behavior to create solutions that fit unique needs. This level of personalization not only improves customer satisfaction but also builds lasting relationships with clients.
- Leading the charge in AI-driven financial solutions are robo-advisors like **Betterment** and **Wealthfront**. These platforms automate investment management, making sophisticated financial strategies accessible to everyday investors. **Robo-advisors** offer personalized investment portfolios that are continuously optimized for performance.
- Meanwhile, JPMorgan's **COiN platform** showcases another facet of AI in finance. COiN streamlines processes that once took thousands of work hours by extracting data from complex legal documents. This efficiency saves time and reduces costs.

AI has immensely impacted finance, transforming everything from fraud prevention to investment strategy and customer engagement. As the industry continues to evolve, AI remains a cornerstone of innovation, driving changes that enhance both the security and personalization of financial services. Whether you're managing investments or simply interacting with your bank, AI is there, quietly revolutionizing how we approach finance.

12.3 AI IN ENTERTAINMENT: CHANGING THE FACE OF MEDIA

Have you ever watched a movie that felt like it was made just for you? That the characters seemed to echo your thoughts, and the plot twists were exactly what you hoped for? This isn't just fantasy; it's the power of AI in entertainment, where content creation and consumption are being revolutionized:

- AI tools streamline production processes, allowing filmmakers to focus on creativity rather than technical tedium. **Automated video editing** software now accelerates post-production, cutting down hours of labor to mere minutes. These tools can analyze footage, make cuts, and even suggest enhancements, making the editor's job easier and more efficient.
- For viewers, AI algorithms suggest movies and shows that match their preferences, analyzing past viewing habits to predict what they might enjoy next, making entertainment more engaging and relevant.
- AI's influence extends into virtual and **augmented reality**, transforming how we interact with digital worlds. Picture a game where the characters respond to your every move, creating a lifelike interaction that blurs the line between reality and fiction. AI-driven real-time character animation makes this possible, allowing game characters to react dynamically based on player inputs. This creates a more immersive experience, drawing players deeper into the narrative.

Interactive storytelling takes it further, where AI creates adaptive narratives that change based on audience choices. Imagine a story that evolves with you, offering different endings and plot paths based on your decisions. This level of interactivity captivates audiences and allows them to explore narratives in new and exciting ways:

- In media marketing, AI optimizes how content creators engage with their audiences. Understanding viewer reactions is crucial, and this is where sentiment analysis comes into play. AI can assess how audiences feel about content, whether they're thrilled, indifferent, or disappointed. This insight helps marketers adapt or change their strategies, ensuring that promotional content resonates with viewers.

- A/B testing, powered by AI, further refines these efforts by enabling creators to experiment with different versions of promotional materials. By analyzing viewer responses, AI identifies which elements are most effective, guiding content creators in maximizing impact and engagement. This data-driven approach allows continuous improvement, ensuring that media campaigns remain fresh and engaging.
- Companies like **LeyLine** have harnessed AI to automate high-end animation production, turning 2D images into 3D assets.
- Meanwhile, **Flawless** uses AI to improve film dubbing, ensuring the dialogue matches the actor's performance naturally. This advancement maintains the integrity of the original acting while enhancing the dubbing process.
- Another example is **Runway**, which offers more than 30 AI-powered creative tools, making advanced video and image creation accessible to content creators. These tools democratize the creative process, allowing more people to bring their ideas to life with the help of AI.

AI in entertainment continues to push boundaries, offering new avenues for creativity, engagement, and innovation. As these technologies evolve, they promise to reshape how we create and consume content, making entertainment more personalized, interactive, and immersive. Whether you're a creator looking to streamline production or a viewer seeking more engaging experiences, AI's impact on the media landscape is undeniable, offering endless possibilities for storytelling and audience engagement.

You've just dived into this rich world of AI and explored its transformational impact across industries like healthcare, finance, and entertainment. Now, it's time to step back and reflect on what you've learned.

12.4 THE ROAD AHEAD: EMBRACING LIFELONG AI LEARNING

To broaden your horizons in AI, consider exploring specialized courses offered by platforms like Coursera and edX. These platforms provide a treasure trove of advanced courses, each covering different aspects of AI:

- **Coursera**, for instance, offers specializations in machine learning and deep learning, catering to those eager to deepen their expertise with courses like "**Advanced Machine Learning Specialization**" and "**Deep Learning Specialization**" by Andrew Ng. These courses don't just teach you technical skills; they also provide insights into real-world applications and the latest research in AI.
- Similarly, **edX** offers university-level courses with certification options, allowing you to gain credentials from prestigious institutions like MIT and Harvard. This is an opportunity to learn from the best and validate your skills with recognized certifications.

In addition to online courses, immersing yourself in the wealth of knowledge in advanced textbooks and publications can significantly deepen your understanding of AI:

- ***Deep Learning* by Ian Goodfellow** is a comprehensive guide that provides an in-depth look at the concepts and methods driving modern AI. It's a valuable resource for anyone looking to grasp the intricacies of deep learning, offering theoretical insights alongside practical applications.
- Moreover, keeping up with journals and conference proceedings like **NeurIPS (Neural Information Processing Systems) and IEEE Transactions on Neural Networks** allows you to stay informed so you can keep up with brealthroughs and cutting-edge research in the field.

These publications are a window into the future of AI, showing innovative ideas and technologies shaping the industry:

- Engage in **coding challenges** and **contribute to open-source projects**. This can be incredibly rewarding for those who prefer to be hands-on. Platforms like **Kaggle** host competitions that challenge you to apply your skills to real-world problems, providing a practical, competitive environment to test and refine your abilities. Participating in these challenges enhances your problem-solving skills and connects you with a community of like-minded individuals passionate about AI.
- **GitHub** is another excellent platform where you can contribute to open-source AI projects. It offers a collaborative space to work on exciting projects, gain practical experience, and learn from others in the community.
- **Podcasts** and **webinars** serve as excellent resources. The **"Artificial Intelligence Podcast" by Lex Fridman** features discussions with leading AI experts, offering insights into the latest trends and technologies. Listen to these conversations and gain a broader perspective on AI's impact across various fields.

Webinars from AI research labs are another way to gain insights into cutting-edge research topics. They offer a platform to learn directly from researchers and industry leaders, informing you about the latest AI developments and innovations.

Interactive Element: Resource List for AI Learning

Coursera: Advanced Machine Learning and Deep Learning Specializations

edX: University-level AI courses with certification options

***Deep Learning* by Ian Goodfellow**

NeurIPS and IEEE Transactions on Neural Networks

Kaggle Competitions: Practical challenges to apply AI skills

GitHub: Contribute to open-source AI projects

"Artificial Intelligence Podcast" by Lex Fridman

AI Research Lab Webinars: Insights into cutting-edge topics

Exploring these resources will enrich your AI learning journey, offering you the knowledge and skills to excel in this dynamic field.

Now that you have all the tools you need to move forward in your AI journey, it's time to share your newfound knowledge and guide other readers toward the same great resources.

By leaving your honest opinion of this book on Amazon, you'll help other beginners in AI discover the information they're looking for and continue their journey of learning AI.

Thank you for your support! The AI community thrives when we share our knowledge—and you're playing a vital role in that.

Click here to leave your review on Amazon.

CONCLUSION

You did it! You've reached the end of our journey together, and I couldn't be more proud of you. When we started this book, my goal was to demystify the world of AI and make it accessible to everyone, regardless of their technical background. I wanted to show you that you don't need to be a coding whiz or have a PhD in computer science to understand and apply AI in your life.

Throughout these pages, we've explored the fundamentals of AI, from its history and core concepts to the practical tools and techniques you can use to build your own AI models.

We've covered topics like choosing the right platform, selecting and preparing datasets, and evaluating and improving your models. We've also explored the ethical considerations surrounding AI and the importance of responsible use.

By now, you should have a solid grasp of the key skills and knowledge needed to navigate AI with confidence. You understand the different types of AI, the basics of machine learning, and how to approach problem-solving with an AI mindset. You've learned how to break down complex tasks into manageable steps and how to leverage the power of AI to enhance your everyday life and catapult your career.

But this is just the beginning of your AI journey. The field of AI is constantly evolving, and there's always more to learn and discover. That's why I encourage you to keep exploring, experimenting, and pushing the boundaries of what's possible with AI.

One of the best ways to continue your learning is to dive into a personal AI project. It doesn't have to be anything fancy or complicated—start small and build from there. Maybe you want to create a chatbot that can answer questions about your favorite hobby or build a model that can predict the weather based on historical data. The possibilities are endless, and building something from scratch will deepen your understanding of AI in ways that reading alone never could.

There are countless online forums, social media groups, and local meetups where you can connect with others who share your passion for AI. These communities are a fantastic resource for asking questions, sharing ideas, and collaborating on projects. You never know what kind of amazing opportunities might arise from these connections.

Don't forget to tap in to the wealth of advanced resources available. From online courses and textbooks to research papers and industry conferences, there's no shortage of ways to deepen your knowledge and stay up-to-date with the latest developments in AI. Set new learning goals for yourself, and don't be afraid to tackle challenging topics—with persistence and practice, you'll be amazed at how far you can go.

Before we part ways, I want to take a moment to thank you for choosing this book as your guide to learning AI. It's been an honor to share my knowledge and experience with you, and I hope you've found our time together both informative and enjoyable. Remember, I'm always here to support you in your continued growth and exploration of AI.

So, go forth and conquer the world of AI! I have no doubt that you have the skills, the knowledge, and the passion to achieve great things. Keep learning, keep exploring, and most importantly, keep having fun. The future of AI is bright, and I can't wait to see what you'll create.

GLOSSARY

- **Abductive Reasoning:** A form of logical inference that seeks the simplest and most likely explanation for observations.
- **Active Learning:** An iterative machine learning technique where the model queries for labels on specific instances to improve its performance.
- **Actor-Critic Method:** A type of reinforcement learning algorithm that combines the benefits of both value-based and policy-based methods.
- **Adversarial Example:** An input to a machine learning model that has been intentionally altered to mislead the model, often used in testing model robustness.
- **Adversarial Training:** A technique that incorporates adversarial examples during the training phase to improve model robustness against deceptive inputs.
- **Agent:** An entity that perceives its environment through sensors and acts upon that environment through actuators.
- **Algorithm:** A finite sequence of well-defined instructions used to solve a problem or perform a task.
- **Artificial General Intelligence (AGI):** Hypothetical AI systems that possess the ability to understand, learn, and apply

intelligence across a wide range of tasks, similar to human cognitive abilities.

- **Artificial Intelligence (AI):** The simulation of human intelligence processes by machines, especially computer systems, including learning, reasoning, and self-correction.
- **Artificial Narrow Intelligence (ANI):** AI systems that are designed and trained for a specific task or narrow domain, as opposed to humanlike general intelligence.
- **Artificial Neural Network (ANN):** A computational model inspired by the way biological neural networks in the human brain work, used in machine learning and pattern recognition.
- **Attention Mechanism:** A method used in models, particularly in natural language processing, that allows the model to focus on specific parts of the input data when making predictions or generating outputs.
- **Backpropagation:** A training algorithm for ANNs that calculates the gradient of the loss function with respect to each weight by the chain rule, allowing the model to update weights efficiently.
- **Bag of Words:** A simplified representation of text data that ignores word order and grammar but retains frequency counts of words.
- **Bagging:** A machine learning ensemble technique that reduces variance by training multiple models (often decision trees) on different subsets of data and aggregating their outputs.
- **Bayesian Inference:** A statistical method that updates the probability for a hypothesis as more evidence or information becomes available, based on Bayes' theorem.
- **Bayesian Network:** A graphical model that represents a set of variables and their conditional dependencies on a directed acyclic graph.
- **Benchmarking:** A process of comparing the performance of various algorithms or models using a standard set of metrics and datasets.
- **Bias-Variance Tradeoff:** A key concept in machine learning that involves balancing the tradeoff between the error due to bias (error from overly simplistic models) and variance (error from overly complex models).

- **Big Data:** Extremely large datasets that may be analyzed computationally to reveal patterns, trends, and associations.
- **Categorical Variable:** A variable that can take on one of a limited, fixed number of possible values, assigning each value to a category.
- **Chatbot:** A software application that uses AI to simulate user conversation, typically through text or voice interactions.
- **Classification:** A supervised learning task that predicts a discrete class label for a given input.
- **Classification Accuracy:** The ratio of correctly predicted instances to the total instances in the dataset, a standard metric for evaluating classification models.
- **Clustering:** An unsupervised learning technique that involves grouping data based on similarities or shared characteristics.
- **Cognitive Computing:** AI systems that mimic human thought processes in complex situations, often designed to enhance human cognitive capabilities.
- **Collaborative Filtering:** A recommendation technique that relies on the past behavior of users to suggest items or information based on similar preferences among users.
- **Computer Vision:** A field of AI that trains computers to interpret and understand the visual world from images or video.
- **Conditional Probability:** The probability of an event occurring given that another event has occurred.
- **Confusion Matrix:** A table used to evaluate the performance of a classification model, showing the number of true positive, true negative, false positive, and false negative predictions.
- **Convolution:** A mathematical operation used in convolutional neural networks that allows the network to learn spatial hierarchies.
- **Convolutional Neural Network (CNN):** A class of deep neural networks primarily used for processing structured grid data, such as images.
- **Cost Function:** A function used to evaluate how well a specific algorithm models the given data, typically representing the difference between predicted and actual values.

- **Data Augmentation:** Techniques used to increase the diversity of training data without collecting new data by applying transformations like rotation, scaling, or flipping to existing data.
- **Data Mining:** The process of discovering patterns and knowledge from large amounts of data.
- **Deep Learning:** A subset of machine learning involving neural networks with many layers (deep neural networks) that can learn intricate patterns in large amounts of data.
- **Deep Reinforcement Learning:** A combination of deep learning and reinforcement learning where deep neural networks are used to make decisions in environments based on rewards and penalties.
- **Dimensionality Curse:** The challenges faced in model training and evaluation when dealing with high-dimensional data, which can lead to overfitting and increased computational costs.
- **Dimensionality Reduction:** The process of reducing the number of features or variables in a dataset while preserving its essential information, often used to simplify models and reduce computation.
- **Ensemble Learning:** A technique that combines multiple models to improve accuracy and robustness, often by leveraging the strengths of each model.
- **Epoch:** One complete pass through the entire training dataset during the training process of a machine learning model.
- **Evolutionary Algorithms:** Computational algorithms inspired by natural evolution, such as genetic algorithms, are used to solve optimization problems.
- **Expert System:** A computer system that emulates the decision-making abilities of a human expert in a specific domain.
- **Explainable Artificial Intelligence (XAI):** Techniques used to make the decisions made by AI systems interpretable and understandable to humans.
- **Exploratory Data Analysis (EDA):** Analyzing datasets to summarize their main characteristics, often using visual methods.
- **Exponential Smoothing:** A statistical technique used to make forecasts or predictions based on past data, assigning exponentially decreasing weights to older data.

- **Feature:** An individual measurable property or characteristic of an observed phenomenon, often used as input in machine learning models.
- **Feature Engineering:** The process of selecting, modifying, or creating features to improve the performance of machine learning models.
- **Feature Importance:** A score that indicates the significance of a feature in predicting the target outcome, often used to interpret models.
- **Feature Vector:** A numerical representation of an object's features, often used as input for machine learning models.
- **Federated Learning:** A distributed machine learning approach where models are trained on decentralized data residing on edge devices while maintaining data privacy.
- **Fully Connected Layer:** A layer in a neural network where each neuron is connected to every neuron in the next layer.
- **Fuzzy Logic:** A form of many-valued logic that deals with reasoning that is approximate rather than fixed and exact, allowing for a range of values as opposed to binary.
- **Generative Adversarial Network (GAN):** A class of machine learning frameworks where two neural networks compete, generating new data with the same statistics as the training dataset.
- **Generative Model:** A type of model that can generate new data points by learning the underlying distribution of the input data (e.g., GANs).
- **Goal-Oriented Agent:** An agent designed to achieve specific objectives and tasks.
- **Gradient Boosting:** An ensemble technique that builds models sequentially, with each new model correcting errors made by the previous ones.
- **Gradient Descent:** An optimization algorithm used to minimize a function iteratively by updating parameters in the direction of the steepest descent of the function.
- **Heuristic:** A practical approach to problem-solving that employs a method not guaranteed to be optimal but sufficient for reaching an immediate goal.

- **Human-in-the-Loop:** A type of machine learning where human feedback is incorporated into the training process to improve model performance.
- **Hyperparameter**: A parameter that is set before the learning process begins and controls the training behavior of the model (e.g., learning rate, batch size).
- **Hyperparameter Optimization:** The process of searching for the most effective hyperparameters for a machine learning model, often utilizing techniques like grid search or random search.
- **Hyperparameter Tuning**: The process of adjusting the hyperparameters of a model to optimize its performance on a given task.
- **Imbalanced Data**: A situation in machine learning where some classes of the data are significantly underrepresented compared to others.
- **Inductive Bias:** Assumptions made by a learning algorithm to predict outputs for inputs that it has not previously encountered.
- **Inference:** The process of drawing conclusions from premises or evidence, often used in AI to derive outputs from model-given inputs.
- **Inference Engine:** The part of an AI system that applies logical rules to the knowledge base to deduce new information or make decisions.
- **Intelligent Agent:** Any entity that perceives its environment and takes actions to maximize its chances of success or reach a goal.
- **Interpretable AI:** AI systems designed to be easily understood by humans, allowing users to comprehend the decision-making processes behind the models.
- **Intuitionistic Logic:** A type of logic that emphasizes constructive proofs, relevant in certain AI applications like reasoning systems.
- **Imputation:** The process of estimating missing values in a dataset to enable complete data utilization in learning processes.
- **Job Scheduler:** A system component that handles the scheduling of jobs in parallel and distributed computing to improve efficiency and resource utilization.
- **Join:** A database operation that combines records from two or more datasets based on a related field.

- **Joint Distribution:** The probability distribution that defines the likelihood of two variables occurring simultaneously.
- **Joint Probability:** The probability of two events happening simultaneously, often used in probabilistic models.
- **Joint Probability Distribution:** A probability distribution that captures the likelihood of two or more random variables occurring simultaneously.
- **K-Fold Cross-Validation:** A method used to assess the performance of a model by dividing the dataset into *k* subsets, training the model *k* times using k-1 subsets for training and one subset for validation.
- **Knowledge Base:** A technology used to store complex structured and unstructured information used by a computer system for decision support.
- **Knowledge Representation:** Techniques used to represent information about the world in a format that a computer system can utilize to solve complex tasks.
- **Labeling:** The process of assigning categorical labels to data points, essential for supervised learning tasks.
- **Latent Dirichlet Allocation (LDA):** A generative statistical model used for topic modeling in document collections, helping to discover underlying themes in text data.
- **Latent Space:** A representation of compressed data, often learned through models like autoencoders, that captures the underlying structure of the data.
- **Latent Variable:** A variable that is not directly observed but is inferred from the observed data, often used in modeling complex phenomena.
- **Likelihood:** A measure of how well a statistical model explains the observed data, often used in parameter estimation.
- **Loss Function:** A function that measures the difference between the predicted output and the actual output during the training of machine learning models.
- **Machine Learning (ML):** A subset of AI that allows systems to learn and improve from experience without being explicitly programmed.

- **Markov Decision Process (MDP):** A mathematical framework for modeling decision-making situations where outcomes are partly random and partly under the control of a decision-maker.
- **Meta-Learning:** A subfield of machine learning focused on learning how to learn; involves algorithms that adapt based on previous experiences.
- **Meta-Modeling:** The use of models to describe models, often used in various AI and machine learning applications to improve system design and analysis.
- **Model:** A mathematical representation of a process that can make predictions or decisions based on input data.
- **Multi-Agent System:** A system composed of multiple interacting agents that can communicate and collaborate to achieve individual or collective goals.
- **Multimodal AI:** AI systems that can process and integrate information from multiple modalities, such as text, images, and audio.
- **Natural Language Generation (NLG):** The process by which AI systems generate human language text based on data inputs.
- **Natural Language Processing (NLP):** A field of AI that focuses on the interaction between computers and humans through natural language.
- **Neural Network:** A series of algorithms that attempt to recognize underlying relationships in a dataset through a process that mimics how the human brain operates.
- **Neural Architecture:** The design and structure of a neural network, including the number and arrangement of layers and the types of functions used.
- **Neural Turing Machine (NTM):** A type of neural network that includes an external memory component, allowing it to store and retrieve information much like a traditional Turing machine.
- **Noise:** Random variability in data that does not represent actual values; can interfere with the learning process in models.
- **Normalization:** The process of adjusting values in datasets to a common scale, often improving the performance of machine learning algorithms by reducing training time and enhancing convergence.

- **Object Detection:** The computer vision task of locating and classifying multiple objects within an image.
- **One-Hot Encoding:** A method of representing categorical variables as binary vectors, where each category is represented by a 1 in its corresponding position and 0s elsewhere.
- **One-shot Learning:** A learning method in which a model learns to recognize patterns from only one or a few examples.
- **Out-of-Distribution (OOD) Samples:** Data points that are drawn from a different distribution than the training data, often posing challenges for model performance.
- **Overfitting:** A modeling error that occurs when a machine learning algorithm captures noise instead of the underlying data distribution, performing well on training data but poorly on unseen data.
- **Overparameterization:** A situation where a model has more parameters than necessary, which can lead to overfitting but may also benefit generalization in specific contexts like deep learning.
- **Perceptron:** The simplest type of artificial neural network used for binary classifications, consisting of input values and weights combined with an activation function.
- **Pipeline:** A sequence of data processing steps in a machine learning workflow, from data collection to model training and evaluation.
- **Policy:** In reinforcement learning, a strategy used by an agent to determine the next action based on the current state.
- **Precision:** A measure of the accuracy of a classifier, defined as the ratio of true positive predictions to the total positive predictions made by the model.
- **Predictive Analytics:** Techniques that use statistical algorithms and machine learning to identify the likelihood of future outcomes based on historical data.
- **Predicate Logic:** A formal logic that uses predicates to express statements about entities; commonly used in knowledge representation in AI.
- **Predictive Model:** A model built using statistical or machine learning techniques to forecast future outcomes based on historical data.

- **Predictive Modeling:** The use of statistics and machine learning techniques to predict future outcomes based on historical data.
- **Processing Time:** The amount of time taken by an algorithm to process an input and produce an output, often a critical factor in real-time systems.
- **Prototype:** A representative example of a category used in classification tasks.
- **Q-Learning:** A reinforcement learning algorithm that seeks to learn the value of actions in particular states to inform future decisions.
- **Quality Assurance:** Processes and measures in AI that ensure models perform as expected, including validation and testing of models against benchmarks.
- **Recall:** A performance metric for classification tasks measuring the proportion of actual positives that were correctly identified by the model (true positives).
- **Recommender System:** An AI system designed to suggest products, services, or content to users based on their preferences and behavior patterns.
- **Recurrent Neural Network (RNN):** A type of neural network designed for processing sequences of data, where connections between nodes can create cycles for feedback loops.
- **Reinforcement Learning:** A type of machine learning where an agent learns to make decisions by taking actions in an environment to maximize cumulative reward.
- **Regression:** A type of predictive modeling technique that estimates relationships among variables, often used to predict continuous outcomes.
- **Regularization:** Techniques used to prevent overfitting by penalizing larger weights or complex models.
- **Reward Function:** A function that defines the goal in reinforcement learning, providing feedback to the agent about the success of its actions.
- **Robotics:** The branch of technology that involves the design, construction, and operation of robots, often incorporating AI to perform tasks autonomously.

- **Scikit-Learn:** An open-source machine learning library for Python that provides simple and efficient tools for data mining and data analysis.
- **Semi-supervised Learning:** A machine learning technique that uses a small amount of labeled data alongside a large amount of unlabeled data to improve learning accuracy.
- **Skip-Gram:** A model used in natural language processing to predict context words from a target word, widely used in Word2Vec.
- **Stochastic Gradient Descent (SGD):** A stochastic approximation of the gradient descent optimization method used in training machine learning models.
- **Supervised Learning:** A type of machine learning where a model is trained on labeled data, meaning the output for each input is known.
- **Support Vector Machine (SVM):** A supervised learning algorithm used for classification and regression tasks that finds the optimal hyperplane to separate data classes.
- **Synthetic Data:** Artificially generated data that mimics the characteristics of real data, often used in training models when real data is scarce or sensitive.
- **Thompson Sampling:** A method for handling the exploration-exploitation dilemma in multi-armed bandit problems, balancing the tradeoff between sampling new actions and exploiting known rewards.
- **Tokenization:** The process of splitting text into individual terms or tokens, crucial in NLP and text analysis tasks.
- **Training Data:** A dataset used to train a model, providing the necessary input-output pairs for the learning process.
- **Training Set:** A subset of the dataset used to train a machine learning model.
- **Transfer Learning:** A technique in machine learning wherein a model developed for one task is reused as the starting point for a model on a second related task.
- **Transformer:** A model architecture that relies on self-attention mechanisms, significantly improving performance in natural language processing tasks.

- **Turing Test:** A measure of a machine's ability to exhibit intelligent behavior indistinguishable from that of a human.
- **Uncertainty Estimation:** The process of quantifying the confidence in model predictions, often crucial in decision-making applications.
- **Underfitting:** A modeling error that occurs when a model is too simple to capture the underlying trend in the data, resulting in poor performance on both training and unseen data.
- **Unsupervised Learning:** A type of machine learning where a model is trained on unlabeled data, meaning no output labels are provided, and the model seeks to learn the underlying structure of the input data.
- **Validation Set:** A subset of data used during the training process to provide an unbiased evaluation of the model's performance while tuning the model's hyperparameters.
- **Variance:** A measure of the variation of a model's predictions; high variance can lead to overfitting, while low variance may cause underfitting.
- **Variational Autoencoder (VAE):** A type of generative model that learns to encode input data into a latent space and then decodes it back to reconstruct the original input, often used for generating new data samples.
- **Variational Inference:** A method in Bayesian statistics used to approximate complex posterior distributions through optimization techniques.
- **Vision-Based Systems:** AI systems that use computer vision techniques to interpret and act upon visual data from the world.
- **Vision Transformer (ViT):** A neural network architecture based on transformers, originally designed for natural language processing, adapted for vision tasks.
- **Weight:** A parameter in a neural network that adjusts the input's influence on the output, typically learned during the training process.
- **Weights Initialization:** The strategy for setting the initial weights of a neural network before training, which is crucial for effective learning.

- **Weight Sharing:** A technique in neural networks where weights are shared across multiple connections to reduce the number of parameters and improve generalization.
- **Word Embedding:** A type of word representation that captures the contextual meaning of words in a dense vector space, allowing for better understanding and similarity measurement of words.
- **XGBoost:** An optimized gradient-boosting library designed to enhance the performance of machine learning models, particularly in structured data tasks.
- **YAML (YAML Ain't Markup Language):** A human-readable data serialization format often used for configuration files and data exchange, especially in machine learning applications.
- **Yardstick:** A term often used in AI to describe a standard measurement or benchmark for evaluating the performance of a model, though not frequently used.
- **Yield Optimization:** A concept in machine learning that focuses on maximizing the output of a process, often used in industries such as agriculture or finance.
- **Yield Prediction:** The use of AI models to predict the output of agricultural crops based on various environmental and management factors.
- **Yottabyte:** A unit of digital information storage equal to one septillion bytes, relevant for discussing the vast amounts of data processed in AI.
- **Zeta Function:** In the context of AI and statistics, a function often involved in the study of complex relationships among variables, though less common than other terms.
- **Zero-Shot Learning:** A type of machine learning wherein a model can perform tasks or classify objects it has never seen before by leveraging knowledge from similar tasks or categories.

This glossary is a foundational resource for understanding key concepts and terminology used in AI. As the field evolves, new terms may emerge, and existing definitions may be refined.

Interactive Element: Flashcard Exercise

- Create a set of digital flashcards using an app like Anki or Quizlet.

- Start with foundational terms like "supervised learning," "deep learning," and "neural networks."
- Test yourself regularly, and challenge yourself to use these terms in conversations or writing to solidify your understanding.

This exercise will help make these concepts feel like second nature, much like learning a new language.

REFERENCES

1. Timeline of artificial intelligence. (n.d.). Wikipedia. https://en.wikipedia.org/wiki/Timeline_of_artificial_intelligence
2. Marr, B. (2023, May 10). 15 amazing real-world applications of AI everyone should know about. Forbes. https://www.forbes.com/sites/bernardmarr/2023/05/10/15-amazing-real-world-applications-of-ai-everyone-should-know-about/
3. AI vs. machine learning: How do they differ? (n.d.). Google Cloud. https://cloud.google.com/learn/artificial-intelligence-vs-machine-learning#:~:text=While%20artificial%20intelligence%20encompasses%20the,accurate%20results%20by%20identifying%20patterns.
4. Kesari, G. (2024, March 31). The future of farming: AI innovations that are transforming agriculture. Forbes. https://www.forbes.com/sites/ganeskesari/2024/03/31/the-future-of-farming-ai-innovations-that-are-transforming-agriculture/#:~:text=The%20AI%20in%20Agriculture%20Market,them%20using%20real%2Dworld%20examples.
5. Beginner-friendly AI platforms GitHub 2023. (2023). Restackio. https://www.restack.io/p/low-code-development-for-ai-answer-beginner-friendly-ai-platforms-github-2023-cat-ai
6. Comparing Google Colab and Microsoft Azure in terms of GPU resources. (n.d.). Medium. https://surtantheta.medium.com/comparing-google-colab-and-microsoft-azure-in-terms-of-gpu-resources-e1fbda8e2ec0
7. Getting started with Python for AI: A complete beginner's guide. (2024, November). Flexiana. https://flexiana.com/2024/11/getting-started-with-python-for-ai-a-complete-beginners-guide/
8. The lazy data scientist's guide to AI/ML troubleshooting. (n.d.). Medium. https://medium.com/@ODSC/the-lazy-data-scientists-guide-to-ai-ml-troubleshooting-abaf20479317
9. The dos and don'ts of dataset selection for machine learning. (n.d.). Medium. https://medium.com/@jan_marcel_kezmann/the-dos-and-donts-of-dataset-selection-for-machine-learning-you-have-to-be-aware-of-8b14513d94a
10. 10 great places to find free datasets for your next project. (n.d.). CareerFoundry. https://careerfoundry.com/blog/data-analytics/where-to-find-free-datasets/
11. AI data privacy challenges and best practices. (n.d.). Granica. https://granica.ai/blog/ai-data-privacy-grc
12. What is data cleaning? Step-by-step guide. (n.d.). Amplitude. https://amplitude.com/explore/data/data-cleaning-guide
13. 23 best data visualization tools you can't miss! (n.d.). Simplilearn. https://www.simplilearn.com/data-visualization-tools-article
14. Python libraries for data clean-up. (n.d.). StrataScratch. https://www.stratascratch.com/blog/python-libraries-for-data-clean-up/

15. Data preprocessing techniques and steps. (n.d.). MATLAB. https://www.mathworks.com/discovery/data-preprocessing.html
16. Create a data integration strategy in 10 steps. (n.d.). Datamation. https://www.datamation.com/big-data/how-to-create-a-data-integration-strategy/
17. The guide to understanding and using AI models. (2024). Viso.ai. https://viso.ai/deep-learning/ml-ai-models/
18. A Gentle Introduction to k-fold Cross-Validation. (n.d.). Machine Learning Mastery. https://www.machinelearningmastery.com/k-fold-cross-validation/
19. How to build an AI model from scratch: A detailed guide. (n.d.). Moon Technolabs. https://www.moontechnolabs.com/blog/how-to-build-an-ai-model/
20. Supervised vs. unsupervised learning: What's the difference? (n.d.). IBM. https://www.ibm.com/think/topics/supervised-vs-unsupervised-learning
21. 12 important model evaluation metrics for machine learning. (2019, August 11). Analytics Vidhya. https://www.analyticsvidhya.com/blog/2019/08/11-important-model-evaluation-error-metrics/#:~:text=Evaluation%20metrics%20are%20quantitative%20measures,comparing%20different%20models%20or%20algorithms.
22. The iterative process of machine learning system. (n.d.). Medium. https://medium.com/@theartifact/the-iterative-process-of-machine-learning-building-and-refining-ml-systems-7ef5aebb7970
23. 8 simple techniques to prevent overfitting. (n.d.). Medium. https://medium.com/towards-data-science/8-simple-techniques-to-prevent-overfitting-4d443da2ef7d
24. How to create effective self-assessments in eLearning: A guide. (n.d.). Create AI Course. https://createaicourse.com/how-to-create-effective-self-assessments-in-elearning/
25. Machine learning models for precise predictive analytics. (n.d.). Stefanini. https://stefanini.com/en/insights/news/machine-learning-models-for-precise-predictive-analytics
26. For retail forecasting, high accuracy predictions in real time. (n.d.). Google Cloud. https://cloud.google.com/blog/topics/retail/retail-forecasting-ai-real-time-insights-vertex-ai-forecast
27. The best AI productivity tools in 2025. (n.d.). Webex Blog. https://blog.webex.com/innovation-ai/the-ai-productivity-tools-shaping-2025/
28. AI-generated art: How can visual artists navigate this? (n.d.). Art Marketing News. https://artmarketingnews.com/ai-generated-art/
29. Saving machine learning models for future use. (n.d.). Medium. https://medium.com/@mohsin.shaikh324/a-comprehensive-guide-saving-machine-learning-models-for-future-use-26777e084100
30. Free AI project hosting platforms. (n.d.). Restackio. https://www.restack.io/p/ai-project-hosting-answer-free-ai-platforms-cat-ai
31. 6 ways to use AI in collaborative tools for smarter project management. (n.d.). Klaxoon. https://klaxoon.com/insight/6-ways-to-use-ai-in-collaborative-tools-for-smarter-project-management
32. Real-world case studies of human-AI collaboration. (n.d.). Smythos. https://smythos.com/ai-agents/agent-architectures/human-ai-collaboration-case-studies/

33. Ethical AI: IEEE's ethically aligned design principles. (n.d.). Wagle, P. https://paulwagle.com/ethical-ai-ieees-ethically-aligned-design-principles/
34. Asilomar AI principles. (n.d.). Future of Life Institute. https://futureoflife.org/open-letter/ai-principles/
35. Addressing data privacy concerns in generative AI. (n.d.). EisnerAmper. https://www.eisneramper.com/insights/artificial-intelligence-insights/data-privacy-concerns-in-generative-ai-1224/
36. Algorithmic bias detection and mitigation: Best practices and policies to reduce consumer harms. (n.d.). Brookings. https://www.brookings.edu/articles/algorithmic-bias-detection-and-mitigation-best-practices-and-policies-to-reduce-consumer-harms/
37. Neural networks explained: Key AI basics. (n.d.). NY Tech Online. https://online.nyit.edu/blog/neural-networks-101-understanding-the-basics-of-key-ai-technology
38. Top 25 deep learning applications used across industries. (n.d.). Simplilearn. https://www.simplilearn.com/tutorials/deep-learning-tutorial/deep-learning-applications
39. Differences between GPT and BERT. (n.d.). GeeksforGeeks. https://www.geeksforgeeks.org/differences-between-gpt-and-bert/#:~:text=While%20both%20GPT%20and%20BERT,deep%20understanding%20of%20language%20context.
40. Five trends in AI and data science for 2025. (n.d.). MIT Sloan Management Review. https://sloanreview.mit.edu/article/five-trends-in-ai-and-data-science-for-2025/
41. How to make an AI model: A step-by-step guide for beginners. (n.d.). Netguru. https://www.netguru.com/blog/how-to-make-an-ai-model
42. The impact of hands-on learning in AI development. (n.d.). AI Startup Connect. https://aistartupconnect.org/blog/the-impact-of-hands-on-learning-in-ai-development
43. How students' self-assessment behavior affects their online learning performance. (n.d.). ScienceDirect. https://www.sciencedirect.com/science/article/pii/S2666920X22000133
44. AI and project-based learning. (n.d.). Project Pals. https://projectpals.com/ai-and-project-based-learning/
45. 7 life-saving AI use cases in healthcare. (n.d.). V7 Labs. https://www.v7labs.com/blog/ai-in-healthcare
46. How artificial intelligence is reshaping the financial services industry. (n.d.). EY. https://www.ey.com/en_gr/insights/financial-services/how-artificial-intelligence-is-reshaping-the-financial-services-industry
47. Husain, A. (2024, August 5). AI takes center stage in media production. Forbes. https://www.forbes.com/sites/amirhusain/2024/08/05/ai-takes-center-stage-in-media-production/
48. AI in media and entertainment: Use cases, benefits, and challenges. (n.d.). LeewayHertz. https://www.leewayhertz.com/ai-in-media-and-entertainment/
49. Artificial intelligence (AI) terms: A to Z glossary. (n.d.). Coursera. https://www.coursera.org/resources/ai-terms

50. Applying AI's potential for motivating and enhancing. (n.d.). SpringerLink. https://link.springer.com/article/10.1007/s10755-024-09747-z
51. Top AI communities in 2023. (n.d.). Medium. https://medium.com/ai-vanguard/top-ai-communities-in-2023-403254cb05f1
52. Create effective quizzes using AI technology. (n.d.). Atlas Blog. https://www.atlas.org/blog/quiz-creation/create-effective-quizzes-using-ai-technology
53. Best AI courses & certificates [2025]: Artificial intelligence. (n.d.). Coursera. https://www.coursera.org/courses?query=artificial%20intelligence
54. Top AI conferences 2023: Roundup of top AI conferences. (2023, February 20). Forbes. https://www.forbes.com/sites/qai/2023/02/20/top-ai-conferences-2023-roundup-of-top-ai-conferences/
55. 12 AI podcasts that will boost your understanding of AI. (n.d.). DigitalOcean. https://www.digitalocean.com/resources/articles/ai-podcasts
56. The impact of AI in community management. (n.d.). Community-Led Growth. https://www.communityledgrowth.com/the-impact-of-ai-in-community-management/